This book is dedicated to my children, Jonathan, Philip, and Mia, for the privilege of being able to share my life with them. It is from these relationships that I realized the importance of writing this book.

This book was written for you if you are hoping to have a baby and need an answer to any of the following questions:

- How long should I just "relax" and try to conceive on my own?
- How can I find out *quickly* and *economically* why I am having trouble becoming pregnant?
- How do I increase my probability of achieving a pregnancy?
- Which infertility treatment should I consider?
- Am I too old to conceive?
- Do we have male infertility?
- What options are there after a vasectomy?
- Am I getting the right tests and treatment?
- Do I need *in vitro* fertilization to become pregnant?
- I do not have regular monthly menstrual periods. Can I conceive?
- I have PCO. Do I need to be treated to become pregnant?
- I have endometriosis. What should I do to conceive?
- I have had repeated miscarriages. What should I do to have a baby?
- I have had a tubal ligation. How can I conceive again?
- Both my ovaries have been removed. What can I do to become pregnant?
- My uterus has been removed. How can I have a baby?

FIND THE CAUSE AND TREATMENT FOR YOUR INFERTILITY

A STEP BY STEP SELF-ASSESSMENT GUIDE

Francis Polansky, M.D.

Reproductive Medicine Solutions
Palo Alto, CA
www.ReproductiveMedicineSolutions.com

Copyright © 2010 by Francis Polansky, M.D.

All rights reserved.

No part of this book may be reproduced, stored in a retrieval system, or transmitted in any form or by any means, electronic, mechanical, recording, or otherwise, without the prior written permission from the author.

Fertility Assessment Algorithm™ is a trademark of Francis Polansky, M.D. and may not be used without written permission.

Published in the United States of America

Library of Congress Control Number: 2010904216

ISBN 978-0-578-05348-6

Printed in the United States of America

First edition 2010

The only source of knowledge is experience.

Albert Einstein

ACKNOWLEDGEMENTS

⌘

I am deeply indebted to my mentor Emmet Lamb, M.D. who, during my fellowship in Reproductive Endocrinology and Infertility at Stanford University, instilled in me the rigorous scientific discipline with which he approached infertility testing and treatments. It is this approach that has served me so well in my own practice of medicine.

I also wish to thank all the patients I have treated over the years since they were my most valuable teachers and a source of knowledge without which this book would not have been possible.

AUTHOR'S NOTE

⌘

This book is a practical guide to an efficient and economical way of finding out what you need to do to improve your chances of achieving a pregnancy. The information and recommendations are based on more than 30 years of my professional experience and extensive research.

Early in my professional career, it became clear that nearly all causes of infertility fall into one or more of the following three categories:

1. The production and quality of the sperm
2. The production and quality of the eggs
3. The joining of the eggs and sperm

This observation formed the basis for the development of this guide for an efficient and expeditious fertility investigation. By eliminating unnecessary steps and delays during your fertility investigation and by avoiding suboptimal treatment options, you may be able to save substantial amounts of money. More importantly, you may be able to avoid the infertility emotional "rollercoaster" and prevent the loss of valuable time and thereby preserve your ability to have a biological child.

At the core of this workbook is the Fertility Assessment Algorithm™. By answering several pivotal questions and based on the results of five commonly used tests, you will be guided step-by-step to the most likely diagnosis and your treatment options.

Whether you are just beginning to think that it may be difficult for you to conceive or you are already struggling with infertility, this book was written for you.

This workbook should be used as a supplement to the treatment plan that your physician recommends and is not intended as a replacement for sound medical advice from a physician. Under no circumstances, should you attempt to use this guide without a physician's supervision or against your physician's advice. On the contrary, sharing of the information contained in this book with your physician is highly recommended.

In the view of ongoing research and the constant flow of information related to reproductive medicine, the author and the publisher make no representations or warranties with respect to the accuracy or completeness of the contents of this work.

All recommendations and procedures contained herein are made without guarantee on the part of the author or the publisher, their agents, or employees. The author and publisher disclaim all liability in connection with the use of the information presented herein.

CONTENTS

⌘

Introduction	1

SECTION I – Your Fertility Potential

How Fertile Are You?	5
Spontaneous Conception	5
Causes of Infertility	6
Male Infertility	7
Ovulation and Egg Quality Factor	7
Aging and Female Fertility Potential	8
The Passage Factor	10
Causes of Infertility: Conclusion	11
Minimizing Delays	11

SECTION II – Optimize Your Fertility

How to Optimize Your Fertility Potential	15

SECTION III – Fertility Assessment Algorithm

Tests Used In the Fertility Assessment Algorithm	21
How to Use the Fertility Assessment Algorithm	21
Fertility Assessment Algorithm	22

SECTION IV – Contemporary Reproductive Treatments

Ovarian Stimulation with Injectable Medications and Intrauterine Insemination — 103

In Vitro Fertilization — 105

Oocyte Donation — 108

Gestational Surrogacy — 112

Oocyte Donation with Gestational Surrogacy — 116

Testicular and Epididymal Sperm Aspiration — 120

Family Gender Balancing and Pre-Implantation Genetic Diagnosis — 121

APPENDIX

Calcium Channel Blockers — 125

About the Author — 126

Index — 128

INTRODUCTION

⌘

A Letter to My Readers

Dear Reader,

Recognizing that you might be infertile can be a shocking realization. Most of us grow up unaware that one day we may face a fifteen percent probability of infertility. That is one in seven people.

The psychological impact of not being able to conceive spontaneously and quickly can be profound. It is not uncommon to experience all the stages of loss: refusal to believe, anger, bargaining, despair and acceptance. Perhaps surprisingly, this feeling of loss seems the same whether you are trying to conceive for the first time or you already have a child or children.

The inability to conceive can also put tremendous stress on a couple's intimate relationship. For many couples, experiencing long-term infertility can be a real "trial-by-fire".

And yet, there can be a "silver lining" in all this anguish: the greater the depth of despair, the higher the elation when you succeed and finally hold your little miracle baby in your arms.

The complexity of infertility diagnosis can be daunting and the cost of treatment can be staggering. It does not have to be that way. This workbook, although it cannot guarantee that you will become pregnant, will help you maximize the probability of success and minimize the expense and the time it may take to conceive.

I will show you how to eliminate unnecessary tests and inefficient treatments by following a unique step-by-step formula of fertility investigation that has a single goal: a baby for you as quickly as possible.

This guide assumes the reader has no knowledge of reproductive medicine. As you proceed with your fertility investigation, you will be simply instructed what step to take next based on your medical history, test results, and treatment outcomes up to that point.

I sincerely hope that this guide will help make your journey to a successful pregnancy as short, economical, and as enjoyable as possible. And, once your baby is born, the mission of this book has truly been achieved.

With warmest wishes of success,

Francis Polansky

Francis Polansky, M.D.

SECTION I
⌘
YOUR FERTILITY POTENTIAL

YOUR FERTILITY POTENTIAL

HOW FERTILE ARE YOU?

Infertility is classically defined as 12 months of unprotected intercourse without achieving a pregnancy. This definition does not take into account the importance of the individual. It is clearly a different situation if the couple is in their early twenties as opposed to their early forties.

Speed of conception in the general population:

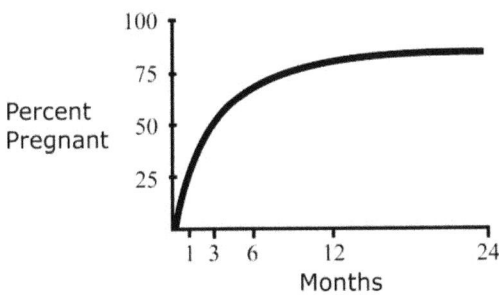

Once a couple has been sexually active without contraception for a year, it is unlikely that they will achieve a spontaneous pregnancy. There is only approximately 1% to 2% probability of a conception during the next ovulation. After two years, the probability approaches 1 in 1000.

Please note that there is no distinction between having unprotected intercourse and "trying" to conceive. They both represent "exposure" to conception.

The likelihood of having a baby depends on your fertility potential as a couple. This potential is a delicate interaction between the male and the female fertility factors. It is believed that in approximately 1/3 of the time, infertility can be solely attributed to a female factor, 1/3 to male factor infertility, and the remaining 1/3 represents a combination of both male and female fertility factors. One can imagine that a very fertile woman could mitigate some degree of her partner's infertility and vice-versa.

Your physical condition could make an enormous difference between successfully achieving a live birth and going through years of frustration. Optimizing your fertility potential can have a considerable impact on the probability of a successful pregnancy[1].

SPONTANEOUS CONCEPTION

Medically unassisted human conception includes the following steps:

1. Ovulation

    Fourteen days prior to the beginning of a new menstrual cycle, an ovarian follicle[2] releases a microscopic egg. Shortly after ovulation, the egg is captured by one of the Fallopian tubes[3] and brought inside the tube.

---

[1] See How to Optimize Your Fertility Potential on page 15.
[2] A grape like structure on the surface of the ovary, almost an inch in diameter.
[3] The connection between the ovaries and the uterus. Fertilization of the eggs takes place within the Fallopian tubes.

2. Fertilization

    Sperm, which can wait in the Fallopian tubes for several days, must fertilize the egg within 12 to 16 hours of ovulation.

3. Early embryonic development

    The fertilized egg (embryo) moves through the Fallopian tube and starts to divide the day after fertilization. In two days, it has 4 cells, in three days, 8 cells, and in six days it has over 100 cells.

4. Endometrial lining

    The female hormones estradiol (estrogen, E_2) and progesterone, produced by the ovulating follicle, prepare the lining of the uterus for implantation.

5. Implantation

    Five to seven days after ovulation, the embryo hatches out of its eggshell, implants into the lining of the uterus and starts to produce the pregnancy hormone human chorionic gonadotropin (HCG). HCG can be detected with a blood pregnancy test approximately 9 to 10 days after ovulation. A urine pregnancy test will typically need close to two weeks after ovulation to turn positive.

CAUSES OF INFERTILITY

There are hundreds of possible causes of infertility. They include male and female structural problems, functional disorders, genetic causes of infertility, hormonal imbalances, and immunological aspects of infertility.

The *vast* majority of infertility causes will fall into one or more of the following three categories:

1. Male factor: Sperm production and sperm quality.

2. Ovulation and hormonal production: Egg production, egg quality, and preparation of the uterine lining for implantation.

3. Passage: The joining of sperm and egg in the Fallopian tubes and transport of the fertilized egg into the uterus.

MALE INFERTILITY

Approximately one third of the time, infertility can be solely attributed to male infertility. The diagnosis of male infertility must be made with consideration of the female fertility potential. For most men, there is no completely reliable test to assess male fertility. The diagnosis of male infertility is often facilitated by first excluding all possible female infertility factors.

Fortunately, we can now overcome all but the most severe forms of male infertility. For the most part, whether you succeed in getting pregnant does not depend on the presence or absence of male infertility. Most infertile men will be able to cause a pregnancy with medical help, but typically male fertility potential cannot be increased by taking medications.

Men can be infertile because they do not produce any sperm. Through testing, it is possible to distinguish between men whose testes are incapable of producing live sperm and those men who produce sperm but have a blockage in the sperm delivery system.

Blockages can be treated surgically or, more recently, it is possible to retrieve the sperm from behind the blockage and perform the intracytoplasmic sperm injection (ICSI) procedure[1]. The same technique can be used for men after vasectomy to avoid a vasectomy reversal surgery.

Most infertile men produce sperm yet may not be able to cause a pregnancy through intercourse or artificial insemination. The vast majority of these men will be able to fertilize their partner's eggs with *in vitro* fertilization[2] and other advanced reproductive treatments (oocyte donation[3], gestational surrogacy[4]).

Unlike female fertility potential, male fertility does not significantly decrease during the first five to six decades of man's life.

OVULATION AND EGG QUALITY FACTOR

Most women ovulate (produce eggs) if their menstrual cycles are regular (variation of no more than seven days between the shortest and the longest cycle) and are less than 40 days apart. Women with longer cycles and greater cycle length variation typically do not ovulate (anovulation) or they ovulate irregularly. There is a wide variety of medications to allow these women to produce eggs. Whether you succeed in getting pregnant does not depend on the presence or absence of ovulation in an *untreated* cycle.

---

[1] Intracytoplasmic sperm injection is a very precise micromanipulation procedure in which a single live sperm is deposited directly into the center of a human egg. It was developed to help couples with male factor infertility and couples whose infertility is unexplained. ICSI has allowed infertile men, who otherwise would not have been able to cause a pregnancy in the past, to father children.
[2] See page 89 for a description of *in vitro* fertilization.
[3] See page 92 for a description of oocyte donation.
[4] See page 95 for a description of gestational surrogacy

A woman's egg quality refers to the level of high quality eggs that are genetically and biologically capable of producing a healthy baby. This egg quality factor depends *primarily* on the female's age. The egg quality is the sole most important factor determining the female fertility potential.

There is no known *medical* treatment to improve the quality of eggs. On the other hand, clinical evidence suggests that a person's lifestyle can have a significant impact on their fertility potential[1].

A woman must have biologically perfect eggs in order to achieve a pregnancy that will go on to a live birth. If the ratio of perfect eggs to biologically suboptimal eggs in a woman is significantly diminished, her chance of having her own genetic child is significantly diminished as well.

Ovaries not only produce eggs; they also produce the female reproductive hormones. The most important of these are estradiol (estrogen, E_2) and progesterone. These hormones are secreted by the granulosa cells lining the inside of the ovarian follicles in which the eggs are maturing.

The function of these cells and the production of the female hormones are dependent on the biological health of the egg inside any given follicle. Usually, normal quality eggs result in normal hormonal levels and vice versa.

One of the common diagnoses of female infertility is ovarian dysfunction (inadequate hormonal production) which almost always refers to the quality of the remaining eggs within the ovaries. In other words, one really cannot have perfect egg quality and, at the same time, inadequate hormonal production. Treatments of female infertility due to ovarian dysfunction should be aimed at increasing the likelihood of producing healthy eggs rather than "subsidizing" the hormonal production.

AGING AND FEMALE FERTILITY POTENTIAL

At birth, a newborn girl has approximately two million eggs within her ovaries. By the time she starts ovulating, she has about 400,000 eggs remaining. From that point on, the ovaries lose approximately 30 eggs a day.

During her reproductive years, a woman ovulates only approximately 400 eggs. When there is only a small portion of the eggs remaining, the highest quality eggs have already been lost. This is why conceptions stop *before* menopause starts.

Female fertility begins to decline many years prior to menopause despite continued regular ovulations. The likelihood of a live birth decreases by approximately 10% to 15% of the

---

[1] See How to Optimize Your Fertility Potential on page 15.

remaining probability each year after the age of 30 to 33 and at an even faster rate after the age of 40.

The following graph illustrates the impact of female age on the female fertility potential:

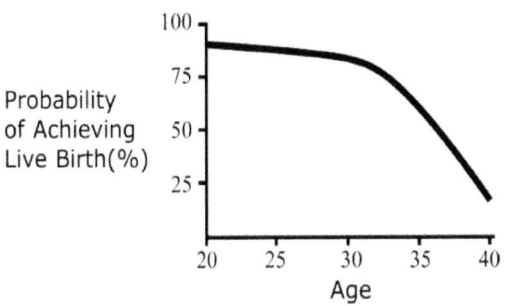

The decrease in female fertility potential is due to the loss of high quality eggs. This age-related loss of fertility magnifies the impact of any other infertility factor present.

Many infertility specialists recommend that infertile women over the age of approximately 38 years should proceed to advanced reproductive treatments quickly before their remaining fertility potential is lost.

As a woman ages, the remaining eggs in her ovaries also age, rendering them less capable of fertilization and of being able to develop into normal embryos. In addition, fertilization of these eggs, whether spontaneous or in a laboratory, is associated with a higher risk of miscarriages and genetic disorders. The vast majority of genetically abnormal pregnancies end very early, often resembling a normal menstrual period.

Risk of chromosomal abnormality in newborns by maternal age:

Maternal Age	Risk of Chromosomal Abnormalities
20	1/526
25	1/476
30	1/385
35	1/192
40	1/66
41	1/53
42	1/42
43	1/33
44	1/26
45	1/21

It is now possible to genetically test early embryos (pre-implantation genetic diagnosis-PGD[1]) as a part of *in vitro* fertilization[2] and other advanced reproductive treatments (oocyte

---

[1] See page 121 for a description of the PGD procedure.
[2] See page 89 for a description of *in vitro* fertilization

donation[1], gestational surrogacy[2]) to minimize the likelihood of transferring genetically abnormal embryos into the uterus.

The likelihood of a fertilized egg implanting is related to the age of the woman who produced the egg and not to the receptivity of the uterus (which does *not* decrease with age). For example, in the oocyte donation treatment, egg donors are typically young women in their twenties, thus the live birth rate for the egg donation treatment varies only slightly regardless of the age of the embryo recipient.

The risk for miscarriage begins to increase among women in their mid-to-late thirties and continues to grow with age, reaching over 40% by the age of 42 years. The miscarriage rates observed among women undergoing advanced reproductive treatments appear to be no higher than in pregnancies conceived through intercourse.

THE PASSAGE FACTOR

In a spontaneous conception, eggs and sperm meet inside the Fallopian tubes. It is quite rare for the sperm not to be able to rapidly arrive in the Fallopian tubes after ejaculation or insemination even if their motility is "sluggish".

The eggs, on the other hand, need at least one healthy Fallopian tube to be captured by after ovulation and brought inside the tube to meet the waiting sperm. There are several conditions (i.e., tubal blockages, pelvic adhesions (scars), adhesions caused by endometriosis[3]) which can make it difficult, if not impossible, for this meeting of the egg and sperm to occur.

It may be possible to repair the damaged Fallopian tubes by operative laparoscopy[4] or laparotomy[5]. Unfortunately, the pregnancy probability after a corrective surgery may be disappointingly low and it may take several years before it is known whether the surgery worked. There is also an increased risk that a pregnancy after a tubal surgery will implant outside the uterus (ectopic pregnancy) possibly requiring additional surgery.

Fortunately, it is possible to bypass the role of the Fallopian tubes altogether with *in vitro* fertilization[6], oocyte (egg) donation[7], or gestational surrogacy[8].

---

[1] See page 92 for a description of oocyte donation
[2] See page 95 for a description of gestational surrogacy
[3] The presence of tissue that normally grows inside the uterus in an abnormal anatomical location.
[4] Surgical procedure during which a thin optical scope is inserted inside the abdomen through a small incision just below the bellybutton. It can be used to correct abnormalities of the reproductive organs.
[5] Major abdominal surgery used to correct abnormalities of the reproductive organs.
[6] See page 89 for a description of *in vitro* fertilization
[7] See page 92 for a description of oocyte donation.
[8] See page 95 for a description of gestational surrogacy.

CAUSES OF INFERTILITY: CONCLUSION

As you can see, the genetic and biological quality of the eggs plays the most important role in infertility treatments. The egg quality, in turn, is very closely tied to the female age.

Where does the receptivity of the mother's body and that of her uterus belong in this scheme? It would seem that uterine lining receptivity should have a profound effect on the success of reproductive treatments. Studies that implicate uterine receptivity as a major factor in infertility treatment success are lacking. On the other hand, there is evidence suggesting that the role of uterine lining receptivity as a decisive factor in human conception is limited.

Since women can have ectopic pregnancies (tubal, ovarian, cervical, abdominal), embryos do not seem to be particularly selective about the environment in which they can implant.

For many years it has been evident that the success of oocyte (egg) donation[1] treatments depends primarily on the donors' egg quality and that the age of the embryo recipient or her uterine receptivity is not an important factor.

The contemporary data suggest that the pregnancy outcome is determined at the moment of the creation of an embryo (the moment of fertilization), and that the subsequent pregnancy events are merely playing out of this predetermined outcome.

To conclude, the embryo quality (egg quality) is the pivotal factor in a successful reproductive outcome.

MINIMIZING DELAYS

An average couple unable to conceive on their own has been through one to two years of infertility. That is one to two years of no contraception and no live birth or ongoing pregnancy, one to two years of testing and treatments, one to two years of heartache and disappointments.

It does not have to be this way. Your infertility investigation can be done systematically and without delays. Within a short period of time, it is possible to arrive at the most appropriate treatment and initiate it. Likewise, even the most complex treatments can be performed in a timely manner.

Once you decide to investigate your fertility potential, there is very little to gain by procrastinating. Yes, there is always the chance of a "miracle" pregnancy but, for some couples, the extra few months of waiting could mean the difference between being able and not being able to have a biological child.

I hope that this fertility guide will encourage you to quickly progress to your ultimate goal: a baby for you.

---

[1] See page 92 for a description of oocyte donation

SECTION II

⌘

OPTIMIZE YOUR FERTILITY

HOW TO OPTIMIZE YOUR FERTILITY POTENTIAL

Your physical condition could make the difference between achieving a live birth and going through years of frustration of unfulfilled dreams. I urge **both partners** to adhere to the following recommendations as closely as feasible and to start implementing them **as soon as possible**.

1. **Optimize your body's acid-alkaline balance**

    The pH of our blood is slightly alkaline. If we eat acidic food, our bodies have to work extra hard to keep the blood in an alkaline state. This extra work stresses our body and can lead to a decrease of one's fertility potential.

    The choices of foods that we eat affect this balance. The typical North American diet is highly acidic. The best way you can maintain the proper blood pH balance is to avoid acid producing foods and increase consumption of alkaline foods. Try not to go out to eat; prepare your own food as much as possible.

    A. Minimize or eliminate the intake of the following acid-forming foods:
    - All grains including corn, oat, and flour-based foods (**breads, pastas, pastry, cereal, dumplings, tortillas, chips...**) except buckwheat and white rice (up to one cooked cup a day)
    - Dairy (**cheese**) except milk, buttermilk, kefir, and yogurt up to one cup a day
    - Alcohol
    - Coffee except de-caffeinated up to two cups a day
    - Cocoa (use carob products instead)
    - Nuts (except hazelnuts and walnuts)
    - Beans/legumes except up to one cup (cooked) a day (not canned)
    - Cranberries (all other berries are ok)
    - **Processed meat** (salami, sausages, hotdogs, canned meat)

    B. Increase intake of the following alkaline foods (organically grown if possible)*:

• Apples	• Cantaloupe	• Grapefruit	• Parsley
• Apricots	• Carrots	• Grapes	• Peach
• Artichoke	• Cauliflower	• **Kale**	• Pear
• Asparagus	• Celery	• Kiwi	• Persimmon
• Avocado	• Chard	• Lemon	• Pineapple
• **Bananas**	• **Coconut**	• Lettuce	• Potatoes

• **Berries**	• Cucumber	• Mango	• **Raisins**
• Beets	• Dates	• Melons (all)	• **Spinach**
• Bell peppers	• Eggplant	• Nectarine	• Salad mix
• Bok choy	• **Figs**	• Olives	• String beans
• Broccoli	• Garlic	• Onions	• Sweet potatoes
• Brussel sprouts	• Ginger	• Orange	• Tomatoes
• Cabbage	• Green peas	• Papaya	• Zucchini

* Items in **bold** are especially helpful.

2. Consume an *abundance* of essential fatty acids:
   - Deep-sea fish and fish oil from non-polluted sources
   - Flaxseed and pumpkin seed oils
   - Broccoli, cauliflower, beets, carrots, kale, collards, cabbage, brussel sprouts
   - Raw seeds
   - Eggs (no more than one a day on average)

3. **Eliminate or minimize intake of trans fatty acids (very important):**
   - Fried foods (if you must have occasional fried food, use coconut oil only)
   - Vegetable shortening
   - Margarine
   - Lard
   - Animal fat
   - Hydrogenated vegetable oils
   - Junk food

4. **Vitamins**

   It is important that **both** partners take high-potency, high-quality natural multivitamins, and mineral supplements of your choice purchased from a reputable source. The female partner must take a minimum of **800 µg of Folic Acid a day**.

5. **Exercise**

   Unless you exercise regularly, several times a week, start *daily* walks (**outdoors**) for a minimum of 45 minutes each day

6. **Volatile Organic Compounds (VOC)**

   Many everyday products release VOC's. It is very important to minimize your exposure (both partners) to VOC's:

- Petroleum products: Avoid car exhaust fumes and solvents, use disposable gloves when filling up your car.
- Release of VOC's from plastics and building materials: Do not drive a new car when trying to conceive, do not remodel your home or buy a newly constructed house.
- No exposure to cigarette smoke (both partners).
- Eliminate or minimize use of perfumes and colognes (unscented deodorant is ok).
- Do not dry-clean your clothes.
- Eliminate air fresheners at home and in your car(s).
- Consider purchasing a VOC-scrubbing air purifier for your bedroom if you sleep with the windows closed (search internet for "voc air purifier").

7. **Fire Retardant Chemicals**

    There is evidence that flame retardants, polybrominated diphenyl ethers (PDBEs), can reduce fertility. Nearly all Americans tested have at least trace levels of flame retardants in their body. Try to minimize your exposure to flame retardants by using bedding and pajamas that are fire retardant free.

8. **Smoking**

    You must not smoke. Cigarette smoking, including passive cigarette smoke exposure, has been shown to have a dramatic adverse effect on oocyte (egg) quality and can also decrease the male fertility potential. Smoking appears to accelerate the loss of eggs and reproductive function and may advance the time of menopause by several years. There is an increased risk of miscarriage and genetic abnormalities in offspring among smokers.

9. **Stress**

    Get plenty of sleep and try to minimize your everyday stresses.

10. **Acupuncture**

    It is ok to have acupuncture

11. **Chinese medicine**

    It is ok to use Chinese herbs as long as they are for strengthening your health only and do not have any male or female hormone-like effect.

SECTION III

⌘

FERTILITY ASSESSMENT ALGORITHM

TESTS USED IN THE FERTILITY ASSESSMENT ALGORITHM

The Fertility Assessment Algorithm™ relies on only five routinely performed reproductive tests:

A. Ovarian Function
    1. Blood level of reproductive hormone follicle stimulating hormone (FSH)
    2. Blood level of reproductive hormone estradiol (estrogen, E_2)
    3. Ultrasound examination of the ovaries
B. Semen Production
    4. Semen analysis (sperm count)
C. Function of the Uterus and Fallopian Tubes
    5. Hysterosalpingogram (HSG, x-ray of the uterus and Fallopian tubes)

Possible additional tests may be required if the HSG findings are abnormal:

- Hysteroscopy: A thin optical scope is passed through the cervical canal inside the uterus and the endometrial cavity is visualized.
- Laparoscopy: A thin optical scope is inserted into the abdomen through a small incision below the bellybutton and the pelvic organs are examined.

HOW TO USE THE FERTILITY ASSESSMENT ALGORITHM

Your fertility investigation can be done systematically and without delays.

Once you start the fertility assessment's step-by-step formula, you will be guided through the few reproductive tests needed to arrive as efficiently as possible at the most likely diagnosis and the corresponding treatment option(s) for you. Which step to take next will depend on the outcome of all the steps you have already taken. Your step sequence will be unique to your particular condition(s).

Since it is not uncommon to have multiple causes of infertility, it is important that you follow the *algorithm's sequence* exactly even if you believe you already know the cause of your infertility.

At the top of each page of this section is a timeline showing the order of progression through the Fertility Assessment Algorithm™. It also shows which segment of the investigation you are currently in.

There is an area at the bottom of each page where you will mark from which page you arrived and to which page you are progressing. **Please fill in the page numbers diligently.** Should there be a need, they will allow you and your physician to retrace the steps of your fertility investigation.

FERTILITY ASSESSMENT ALGORITHM

Preliminary Inquiry → Ovarian Function → Semen → Uterus & Fallopian Tubes → Treatment

FERTILITY ASSESSMENT ALGORITHM

Please choose the **FIRST** statement that applies to you:

Condition	Action
1. *Both* ovaries have been removed.	Go to page 23
2. Your uterus has been removed (hysterectomy).	Go to page 24
3. You were told that, for medical reasons, you cannot or should not carry a pregnancy.	Go to page 25
4. *Both* Fallopian tubes have been removed.	Go to page 26
5. *Both* Fallopian tubes are blocked (including previous tubal ligation).	Go to page 27
6. You (female partner) are 42 years or older.	Go to page 28
7. The *typical* number of days between onsets of your periods is 40 or more.	Go to page 29
8. You do not have menstrual periods.	Go to page 30
9. Because of previous vasectomy, there are no sperm in the semen.	Go to page 31
10. You can conceive *without difficulty* but you have had *three or more* miscarriages.	Go to page 32
11. None of the above statements apply to you.	Go to page 33

Go to page ☐

FERTILITY ASSESSMENT ALGORITHM

Preliminary Inquiry → Ovarian Function → Semen → Uterus & Fallopian Tubes → Treatment

YOUR OVARIES HAVE BEEN REMOVED

If both ovaries have been removed (oopherectomy), you will need to use an egg donor to conceive (oocyte donation[1] treatment).

Since the semen quality will not influence the choice of treatment and since a semen examination will be a part of oocyte donation prerequisites, your next step is an assessment of your endometrial cavity (the inside of your uterus) to make sure it can support a pregnancy.

Also, if no sperm are being ejaculated because of the male partner's previous vasectomy, a sperm aspiration[2] procedure will be added to oocyte donation.

For instructions on an endometrial cavity assessment: **Go to page 77.**

If your uterus has been removed as well (hysterectomy), you will need either "traditional" surrogacy or a combination of oocyte donation and "gestational" surrogacy.

In traditional surrogacy, the surrogate will provide both the egg and carry the pregnancy. She conceives through an artificial insemination with the intended father's semen. This is a simple procedure which can be done by your gynecologist. You should not need an infertility specialist for this treatment.

In oocyte donation with gestational surrogacy[3], the egg donor and the surrogate are two different women. The donor's ovaries are stimulated to produce multiple eggs, the intended father's semen is used to fertilize the eggs and the resulting embryo(s) is/are deposited into the surrogate's uterus. The egg donor could be related to the intended mother (sister, cousin, niece, etc.) or she could come from one of the many egg donor agencies.

No further fertility assessment is needed since a semen examination will be a part of prerequisites for oocyte donation with gestational surrogacy.

Also, if no sperm are being ejaculated because of previous vasectomy, a sperm aspiration[2] procedure will be added to oocyte donation with gestational surrogacy.

To find out more about oocyte donation with gestational surrogacy treatment: **Go to page 98.**

---

[1] See page 92 for a description of oocyte donation.
[2] See page 120 for a description of the sperm aspiration procedure.
[3] See page 98 for a description of oocyte donation with gestational surrogacy.

From page ☐   Go to page ☐

FERTILITY ASSESSMENT ALGORITHM

Preliminary Inquiry → Ovarian Function → Semen → Uterus & Fallopian Tubes → Treatment

YOUR UTERUS HAS BEEN REMOVED

Since you still have one or both ovaries, you may be able to conceive with your eggs but you will need to use a surrogate woman to carry your pregnancy. You will need "gestational" surrogacy[1] treatment.

Also, if no sperm are being ejaculated because of the male partner's previous vasectomy, a sperm aspiration[2] procedure will be added to gestational surrogacy.

Your next step is to find out whether you can successfully (pregnancy not ending in a miscarriage) conceive with your eggs; you need to assess your egg quality (ovarian reserve): **Go to page 36.**

---

[1] See page 95 for a description of gestational surrogacy.
[2] See page 120 for a description of the sperm aspiration procedure.

FERTILITY ASSESSMENT ALGORITHM

Preliminary Inquiry → Ovarian Function → Semen → Uterus & Fallopian Tubes → Treatment

YOU WERE TOLD THAT, FOR MEDICAL REASONS, YOU CANNOT OR SHOULD NOT CARRY A PREGNANCY

You will need to use a surrogate woman to carry your pregnancy. You will need "gestational" surrogacy[1] treatment.

Also, if no sperm are being ejaculated because of the male partner's previous vasectomy, a sperm aspiration[2] procedure will be added to gestational surrogacy.

Your next step is to find out whether you can successfully (pregnancy not ending in a miscarriage) conceive with your eggs; you need to assess your egg quality (ovarian reserve): **Go to page 36.**

---

[1] See page 95 for a description of gestational surrogacy.
[2] See page 120 for a description of the sperm aspiration procedure.

From page ☐   Go to page ☐

FERTILITY ASSESSMENT ALGORITHM

Preliminary Inquiry → Ovarian Function → Semen → Uterus & Fallopian Tubes → Treatment

BOTH FALLOPIAN TUBES HAVE BEEN REMOVED

There is no communication between your ovaries and the uterus. The eggs and sperm cannot meet inside your body. You need *in vitro* fertilization (IVF)[1] to bypass the function of the Fallopian tubes.

Also, if no sperm are being ejaculated because of the male partner's previous vasectomy, a sperm aspiration[2] procedure will be added to *in vitro* fertilization.

To determine whether you are a candidate for *in vitro* fertilization, you must first assess your egg quality (ovarian reserve): **Go to page 36.**

---

[1] See page 89 for a description of *in vitro* fertilization.
[2] See page 120 for a description of the sperm aspiration procedure.

From page [ ] Go to page [ ]

FERTILITY ASSESSMENT ALGORITHM

Preliminary Inquiry → Ovarian Function → Semen → Uterus & Fallopian Tubes → Treatment

BOTH FALLOPIAN TUBES ARE BLOCKED

There is no communication between your ovaries and the uterus. The eggs and sperm cannot meet inside your body.

If you had a tubal ligation, it is sometimes possible to undo the sterilization procedure. This typically requires a major abdominal operation and it may be years after the surgery before you would know whether it was successful or not. The repair also carries with it an increased risk of a tubal pregnancy. For these reasons, the tubal re-anastomosis surgery has been gradually replaced with *in vitro* fertilization (IVF)[1].

IVF bypasses the function of the Fallopian tubes and the presence of a tubal blockage does not influence the probability of success.

Also, if no sperm are being ejaculated because of the male partner's previous vasectomy, a sperm aspiration[2] procedure will be added to *in vitro* fertilization.

To determine whether you are a candidate for *in vitro* fertilization, you must first assess your egg quality (ovarian reserve): **Go to page 36.**

---

[1] See page 89 for a description of *in vitro* fertilization.
[2] See page 120 for a description of the sperm aspiration procedure.

From page ☐   Go to page ☐

FERTILITY ASSESSMENT ALGORITHM

Preliminary Inquiry → Ovarian Function → Semen → Uterus & Fallopian Tubes → Treatment

YOU ARE 42 YEARS OR OLDER

If you are younger than 50 years:

For most women, it is unlikely to conceive successfully (pregnancy that does not end in a miscarriage) with their own eggs past the age of 42. For the ones who can, time is of essence. You should, without a delay, move to advanced infertility treatments (i.e., *in vitro* fertilization[1]).

Also, if no sperm are being ejaculated because of the male partner's previous vasectomy, a sperm aspiration[2] procedure will be added to the treatment.

To determine which advanced reproductive treatment is most appropriate for you, your next step is an assessment of your egg quality (ovarian reserve): **Go to page 36.**

If you are 50 years or older:

There is no biological age cut-off point for carrying a pregnancy (from egg donation) but since serious pregnancy-related complications increase with age and could compromise the mother's and baby's health, you should not try to become pregnant if you are older than 50 years.

As an alternative, you might consider "traditional" surrogacy or a combination of egg donation and "gestational" surrogacy.

In traditional surrogacy, the surrogate will both provide the egg and carry the pregnancy. She conceives through an artificial insemination with the intended father's semen. This is a simple procedure which can be done by your gynecologist.

In egg donation with gestational surrogacy, the egg donor and the surrogate are two different women. The donor's ovaries are stimulated to produce multiple eggs, the intended father's semen is used to fertilize the eggs and the resulting embryo(s) is/are deposited into the surrogate's uterus. The egg donor could be related to the intended mother (sister, cousin, niece, etc.) or she could come from one of the many egg donor agencies.

Also, if no sperm are being ejaculated because of the intended father's previous vasectomy, a sperm aspiration[2] procedure will be added to the treatment.

To find out more about the combination of oocyte donation and gestational surrogacy: **Go to page 98.**

---

[1] See page 89 for a description of *in vitro* fertilization
[2] See page 120 for a description of the sperm aspiration procedure.

From page ☐   Go to page ☐

TYPICAL NUMBER OF DAYS BETWEEN ONSETS OF YOUR PERIODS IS 40 OR MORE DAYS

If the onsets of your menstrual periods are 40 or more days apart, you likely do not produce eggs regularly, therefore you do not ovulate (anovulation) or ovulate irregularly.

A lack of regular ovulation is a relatively common condition which is normally easy to diagnose and, for most women, the appropriate treatment can be highly successful.

Also, if no sperm are being ejaculated because of the male partner's previous vasectomy, a sperm aspiration[1] procedure or a microsurgical reversal of vasectomy will be needed.

You should first establish the cause(s) of the lack of regular ovulations; you need to assess your egg quality (ovarian reserve): **Go to page 36.**

---

[1] See page 120 for a description of the sperm aspiration procedure.

FERTILITY ASSESSMENT ALGORITHM

Preliminary Inquiry → Ovarian Function → Semen → Uterus & Fallopian Tubes → Treatment

YOU DO NOT HAVE MENSTRUAL PERIODS

A lack of menstrual periods can be caused by a lack of ovulation (anovulation) and normal ovarian hormones production, the absence of growth of the uterine lining, or an obstruction in the uterine outflow.

Of these causes, the lack of ovulation is the most common one. It is normally easy to diagnose and, for most women, the treatment can be highly successful but may require advance fertility treatment(s).

Also, if no sperm are being ejaculated because of the male partner's previous vasectomy, a sperm aspiration[1] procedure or a microsurgical reversal of vasectomy will be needed.

The first step in your fertility assessment is the determination of the egg quality (ovarian reserve): **Go to page 36.**

---

[1] See page 120 for a description of the sperm aspiration procedure.

From page ☐   Go to page ☐

FERTILITY ASSESSMENT ALGORITHM

| Preliminary Inquiry | → | Ovarian Function | → | Semen | → | Uterus & Fallopian Tubes | → | Treatment |

BECAUSE OF PREVIOUS VASECTOMY, THERE ARE NO SPERM IN THE SEMEN

Vasectomy is a method of male sterilization in which the tubes (vas deferens) carrying sperm from the testicles are cut and tied off.

There are two options of being able to cause a pregnancy after a vasectomy: a sperm aspiration[1] procedure and a microsurgical reversal of the vasectomy.

The decision between these two alternatives is influenced by the quality of the female partner's eggs and the condition of her Fallopian tubes.

The first step in your fertility assessment is the determination of the egg quality (ovarian reserve): **Go to page 36.**

---

[1] See page 120 for a description of the sperm aspiration procedure.

FERTILITY ASSESSMENT ALGORITHM

Preliminary Inquiry → Ovarian Function → Semen → Uterus & Fallopian Tubes → Treatment

YOU HAVE HAD THREE OR MORE MISCARRIAGES

Repeated (recurrent) miscarriages are defined as at least three or more consecutive miscarriages.

Having a miscarriage is a very common occurrence. It is estimated that only one in three or four pregnancies conceived by young, fertile couples will be able to advance to a live birth. Most of the other pregnancies will end in very early miscarriages. By the time a *young* woman starts to miss her menstrual period and has a positive pregnancy test, the risk of a miscarriage has decreased to approximately 25%.

Statistically, having one or two miscarriages should not make it more likely that you will have yet another one. After three miscarriages, the probability of another miscarriage increases only by approximately 5% to 10%. This means that even after several miscarriages, your next pregnancy will likely result in a live birth.

The first step in the recurrent miscarriage investigation is the determination of your egg quality (ovarian reserve): **Go to page 36.**

From page ☐ Go to page ☐

FERTILITY ASSESSMENT ALGORITHM

Preliminary Inquiry → Ovarian Function → Semen → Uterus & Fallopian Tubes → Treatment

LENGTH OF "EXPOSURE" TO CONCEPTION[1]

Let's first address the issue of "length of infertility":

There is no distinction between having unprotected intercourse and "trying" to conceive. They both represent "exposure" to conception. It is very important that you do not wait too long. On the other hand, you and your partner should have sufficiently long "exposure" to pregnancy before spending time, money, and emotional energy on fertility tests and treatment.

In the table below, match the female partner's age and the length of time (in months) that you have been sexually active without doing anything to keep you from conceiving. Note that miscarriages are counted as infertility.

If, on average, you are having intercourse four times a month or more often, you do not have to worry about timing intercourse to ovulation. There is no upper limit on the frequency of intercourse.

This table will guide you when to start your fertility testing. Your first test should be an assessment of the quality of your eggs (your ovarian reserve). Starting with an egg quality assessment is the most efficient way of carrying out your fertility investigation.

Length of Exposure / Female Age	No Exposure	1 to 3 months	4 to 6 months	7 to 9 months	10 to 12 months	13 or more months
Less than 36	Begin sexual activity without contraception	Continue sexual activity without contraception	Continue sexual activity without contraception	Continue sexual activity without contraception	Continue sexual activity without contraception	Go to page 36
36 to 37	Begin sexual activity without contraception	Continue sexual activity without contraception	Continue sexual activity without contraception	Continue sexual activity without contraception	Go to page 36	Go to page 36
38 to 39	Begin sexual activity without contraception	Continue sexual activity without contraception	Continue sexual activity without contraception	Go to page 36	Go to page 36	Go to page 36
40 to 41	Begin sexual activity without contraception	Continue sexual activity without contraception	Go to page 36	Go to page 36	Go to page 36	Go to page 36

[1]The male partner should not be taking calcium channel blocker medication (see page 125).

From page ☐   Go to page ☐

FERTILITY ASSESSMENT ALGORITHM

Preliminary Inquiry → Ovarian Function → Semen → Uterus & Fallopian Tubes → Treatment

LENGTH OF "EXPOSURE" TO CONCEPTION[1]

After you have started ovulating, your "exposure" to conception begins.

There is no distinction between having unprotected intercourse and "trying" to conceive. They both represent "exposure" to conception. It is very important that you do not wait too long. On the other hand, you and your partner should have sufficiently long "exposure" to pregnancy before spending time, money, and emotional energy on additional fertility tests and treatment.

In the table below, match the female partner's age and the length of time (in months) since you started to have regular menstrual cycles. Note that miscarriages are counted as infertility.

If, on average, you are having intercourse four times a month or more often, you do not have to worry about timing intercourse to ovulation. There is no upper limit on the frequency of intercourse.

This table will guide you when to proceed to the assessment of the male partner's semen quality as your next step. Make sure that you continue to have regular periods.

If your cycles again become irregular and/or 40 or more days apart, you will need ovarian stimulation treatment. Your next step then is an assessment of semen quality: **Go to page 53.**

Length of Exposure / Female Age	1 to 3 months	4 to 6 months	7 to 9 months	10 to 12 months	13 or more months
Less than 36	Continue sexual activity without contraception	Continue sexual activity without contraception	Continue sexual activity without contraception	Continue sexual activity without contraception	Go to page 53
36 to 37	Continue sexual activity without contraception	Continue sexual activity without contraception	Continue sexual activity without contraception	Go to page 53	Go to page 53
38 to 39	Continue sexual activity without contraception	Continue sexual activity without contraception	Go to page 53	Go to page 53	Go to page 53
40 to 41	Continue sexual activity without contraception	Go to page 53	Go to page 53	Go to page 53	Go to page 53

[1]The male partner should not be taking calcium channel blocker medication (see page 125).

From page ☐   Go to page ☐

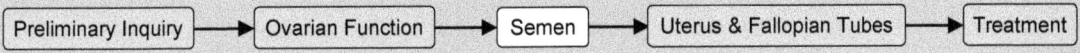

FERTILITY ASSESSMENT ALGORITHM

LENGTH OF "EXPOSURE" TO CONCEPTION[1]

After a successful surgical repair of the blockage in the sperm delivery system, you should give yourselves sufficiently long "exposure" to conception. You do not have to "try" to conceive as long as you are not using contraception. It is also very important that you do not wait too long.

If, on average, you are having intercourse four times a month or more often, you do not have to worry about timing intercourse to ovulation. There is no upper limit on the frequency of intercourse.

In the table below, match the female partner's age and the length of time (in months) after sperm production started following the surgical repair. If you do not conceive during the recommended period of time, it is likely that in spite of the sperm production, the sperm quality is not sufficiently high to cause a pregnancy through intercourse. *In vitro* fertilization[2] (IVF) may then become the most appropriate treatment for you.

Before the IVF treatment, your uterine cavity will need to be checked for presence of endometrial polyps, internal myomas (fibroids, hard nodules inside the uterine cavity), and intrauterine adhesions (scarring).

Length of Exposure / Female Age	1 to 3 months	4 to 6 months	7 to 9 months	10 to 12 months	13 or more months
Less than 36	Continue sexual activity without contraception	Continue sexual activity without contraception	Continue sexual activity without contraception	Continue sexual activity without contraception	Go to page 74
36 to 37	Continue sexual activity without contraception	Continue sexual activity without contraception	Continue sexual activity without contraception	Go to page 74	Go to page 74
38 to 39	Continue sexual activity without contraception	Continue sexual activity without contraception	Go to page 74	Go to page 74	Go to page 74
40 to 41	Continue sexual activity without contraception	Go to page 74	Go to page 74	Go to page 74	Go to page 74

[1] The male partner should not be taking calcium channel blocker medication (see page 125).
[2] See page 89 for a description of *in vitro* fertilization.

From page [ ]   Go to page [ ]

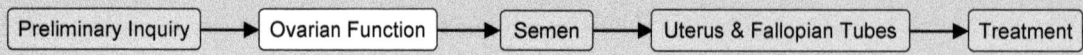

ASSESSMENT OF OVARIAN RESERVE

Since the ability of ovaries to produce normal eggs plays a central role in the process of conception, an assessment of your ovarian function (ovarian reserve testing) should be your first test.

A woman's egg quality refers to the level of high quality eggs that are genetically and biologically capable of producing a healthy baby. Examining your ovarian reserve will require a blood test to measure levels of follicle stimulating hormone (FSH) and estradiol (estrogen, E_2) and an ultrasound examination of your ovaries:

1. Hormonal Assessment:

   FSH, produced by the pituitary gland located at the base of the brain, stimulates ovaries to produce eggs. If the ovaries cannot produce normal eggs, the FSH level is increased. Estradiol production by the ovaries influences the FSH secretion and is also related to the quality of the eggs.

   If you had your FSH and estradiol measured (from the same blood sample) within the last three months, use those results. **Please note that you must not take any hormonal medications for at least four weeks prior to the test.**

   To schedule your FSH and estradiol testing:

Condition	Action
• Your periods are less than 40 days apart.	• Your blood must be drawn on cycle day 2, 3, or 4. Cycle day 1 is the first day of a *full* flow.
• Your periods are 40 or more days apart. • You do not have menstrual periods.	• You can have your blood drawn anytime, regardless of your cycle day.
• You had a hysterectomy.	• You will have to use "ovulation predictor" testing[1] to determine when to have your blood drawn.

   [1] Urine ovulation predictor tests can be purchased over the counter at a drug store. Start your testing any day and continue once a day at approximately the same time. It may take 30 or more days before the test turns positive. Once the test has turned positive, have your blood drawn 16, 17, or 18 days later. If you have tested for 39 consecutive days without the test turning positive, stop the testing and have your blood drawn any day.

   - Continue on the next page -

From page

FERTILITY ASSESSMENT ALGORITHM

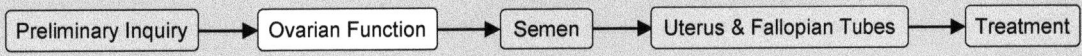

2. The ultrasound of your ovaries will determine the number of antral follicles (see below) within the ovaries. Their number is related to the quality of your eggs. Ideally, there should be 5 to 8 antral follicles **per ovary.**

The ultrasound can be scheduled anytime during your cycle, even when you are having a menstrual period. If you had an ultrasound assessment of the antral follicles within the last six months, you do not have to repeat it.

You will need to emphasize to the person doing the ultrasound that you want to know the **total number of antral follicles in *both* ovaries**. Antral follicles will be seen as dark circles typically found in the periphery of the ovaries.

In the following table, enter your ultrasound and FSH/E_2 results as "Hormonal test #1". If you are later instructed to repeat the hormonal test, use the 2nd and the 3rd row.

	Date	Total number of antral follicles for both ovaries	FSH level	Estradiol (E_2) level
Ultrasound			N/A	N/A
Hormonal test #1		N/A		
Hormonal test #2		N/A		
Hormonal test #3		N/A		

In the ovarian reserve tables on pages 38 through 42, match your total antral follicle count for **both** ovaries with your FSH and estradiol results to find your next step. If one of your ovaries has been surgically removed, multiply the number of antral follicles within the remaining ovary by 2.

Your age will determine which ovarian reserve table to use:

Female Age	Table
Less than 35	Go to page 38
35 to 37	Go to page 39
38 to 40	Go to page 40
41 to 42	Go to page 41
43 and older	Go to page 42

Go to page ☐

FERTILITY ASSESSMENT ALGORITHM

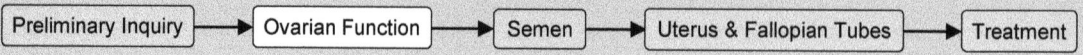

Ovarian Reserve Tables
Female Age Less than 35 Years

0 to 6 Antral Follicles (Total for both ovaries)

E_2 \ FSH	Less than 8.0	8.0 to 9.9	10.0 to 12.9	13.0 or greater
Less than 50	Go to page 43	Go to page 43	Go to page 44	Go to page 44
50 to 69	Go to page 43	Go to page 43	Go to page 44	Go to page 44
70 or greater	Go to page 43	Go to page 44	Go to page 44	Go to page 44

7 or More Antral Follicles (Total for both ovaries)

E_2 \ FSH	Less than 8.0	8.0 to 9.9	10.0 to 12.9	13.0 or greater
Less than 50	Go to page 43	Go to page 43	Go to page 43	Go to page 44
50 to 69	Go to page 43	Go to page 43	Go to page 44	Go to page 44
70 or greater	Go to page 43	Go to page 44	Go to page 44	Go to page 44

From page [ ]   Go to page [ ]

FERTILITY ASSESSMENT ALGORITHM

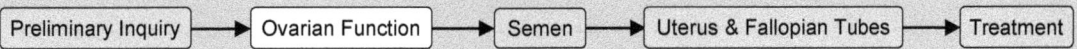

Ovarian Reserve Tables
Female Age 35 to 37 Years

0 to 6 Antral Follicles (Total for both ovaries)

E_2 \ FSH	Less than 8.0	8.0 to 9.9	10.0 to 12.9	13.0 or greater
Less than 50	Go to page 43	Go to page 43	Go to page 44	Go to page 44
50 to 69	Go to page 43	Go to page 44	Go to page 44	Go to page 44
70 or greater	Go to page 43	Go to page 44	Go to page 44	Go to page 46

7 or More Antral Follicles (Total for both ovaries)

E_2 \ FSH	Less than 8.0	8.0 to 9.9	10.0 to 12.9	13.0 or greater
Less than 50	Go to page 43	Go to page 43	Go to page 44	Go to page 44
50 to 69	Go to page 43	Go to page 43	Go to page 44	Go to page 44
70 or greater	Go to page 43	Go to page 44	Go to page 44	Go to page 44

From page [ ]   Go to page [ ]

FERTILITY ASSESSMENT ALGORITHM

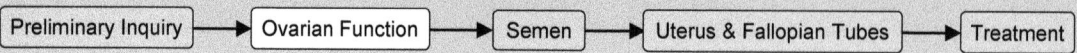

Ovarian Reserve Tables
Female Age 38 to 40 Years

0 to 6 Antral Follicles (Total for both ovaries)

E_2 \ FSH	Less than 8.0	8.0 to 9.9	10.0 to 12.9	13.0 or greater
Less than 50	Go to page 43	Go to page 44	Go to page 44	Go to page 44
50 to 69	Go to page 44	Go to page 44	Go to page 44	Go to page 44
70 or greater	Go to page 44	Go to page 44	Go to page 44	Go to page 46

7 or More Antral Follicles (Total for both ovaries)

E_2 \ FSH	Less than 8.0	8.0 to 9.9	10.0 to 12.9	13.0 or greater
Less than 50	Go to page 43	Go to page 44	Go to page 44	Go to page 44
50 to 69	Go to page 43	Go to page 44	Go to page 44	Go to page 44
70 or greater	Go to page 43	Go to page 44	Go to page 44	Go to page 46

From page [ ] Go to page [ ]

FERTILITY ASSESSMENT ALGORITHM

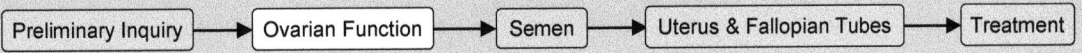

Ovarian Reserve Tables
Female Age 41 to 42 Years

0 to 6 Antral Follicles (Total for both ovaries)

E_2 \ FSH	Less than 8.0	8.0 to 9.9	10.0 to 12.9	13.0 or greater
Less than 50	Go to page 44	Go to page 44	Go to page 44	Go to page 46
50 to 69	Go to page 44	Go to page 44	Go to page 44	Go to page 46
70 or greater	Go to page 44	Go to page 44	Go to page 44	Go to page 46

7 or More Antral Follicles (Total for both ovaries)

E_2 \ FSH	Less than 8.0	8.0 to 9.9	10.0 to 12.9	13.0 or greater
Less than 50	Go to page 44	Go to page 44	Go to page 44	Go to page 44
50 to 69	Go to page 44	Go to page 44	Go to page 44	Go to page 46
70 or greater	Go to page 44	Go to page 44	Go to page 44	Go to page 46

From page ☐   Go to page ☐

FERTILITY ASSESSMENT ALGORITHM

Preliminary Inquiry → Ovarian Function → Semen → Uterus & Fallopian Tubes → Treatment

Ovarian Reserve Tables
Female Age 43 Years and Older

0 to 6 Antral Follicles (Total for both ovaries)

E₂ \ FSH	Less than 8.0	8.0 to 9.9	10.0 to 12.9	13.0 or greater
Less than 50	Go to page 44	Go to page 44	Go to page 46	Go to page 46
50 to 69	Go to page 44	Go to page 44	Go to page 46	Go to page 46
70 or greater	Go to page 44	Go to page 44	Go to page 46	Go to page 46

7 or More Antral Follicles (Total for both ovaries)

E₂ \ FSH	Less than 8.0	8.0 to 9.9	10.0 to 12.9	13.0 or greater
Less than 50	Go to page 44	Go to page 44	Go to page 44	Go to page 46
50 to 69	Go to page 44	Go to page 44	Go to page 44	Go to page 46
70 or greater	Go to page 44	Go to page 44	Go to page 46	Go to page 46

From page ☐   Go to page ☐

FERTILITY ASSESSMENT ALGORITHM

Preliminary Inquiry → Ovarian Function → Semen → Uterus & Fallopian Tubes → Treatment

YOU SHOULD BE ABLE TO CONCEIVE WITH YOUR EGGS

The results of your ovarian reserve testing suggest that you should be able to conceive with your eggs. No additional egg quality testing is necessary at this point.

To find your next step, please choose the **FIRST** statement that applies to you:

Condition	Action
1. You had a hysterectomy. 2. You were told that, for medical reasons, you cannot or should not carry a pregnancy.	• Your fertility assessment is complete; no semen test is needed at this point, since semen examination will be a part of prerequisites for gestational surrogacy. You can proceed to gestational surrogacy: **Go to page 95.**
3. Both Fallopian tubes have been removed. 4. Both Fallopian tubes are blocked.	• Your next step is an assessment of your endometrial cavity. No semen test is needed at this point, since semen examination will be a part of prerequisites for *in vitro* fertilization. For the assessment of your endometrial cavity: **Go to page 74.**
5. You are 42 years or older.	• You are a candidate for *in vitro* fertilization. No semen test is needed at this point, since semen examination will be a part of prerequisites for *in vitro* fertilization. Your next step is an assessment of your endometrial cavity: **Go to page 74.**
6. The typical number of days between onsets of your periods is 40 or more days. 7. You do not have menstrual periods	• If no sperm are being ejaculated because of previous vasectomy: **Go to page 52.** • Otherwise your next step is the investigation of the lack of regular ovulations. **Go to page 47.**
8. None of the above statements apply to you.	• If no sperm are being ejaculated because of previous vasectomy: **Go to page 52.** • Otherwise your next step is an assessment of the semen quality: **Go to page 53.**

From page [ ]   Go to page [ ]

FERTILITY ASSESSMENT ALGORITHM

Preliminary Inquiry → Ovarian Function → Semen → Uterus & Fallopian Tubes → Treatment

YOU NEED TO CONSIDER ADVANCED REPRODUCTIVE TREATMENTS

Your ovarian reserve test results suggest that the quality of your eggs may be insufficient to conceive without the help of one of the advanced reproductive treatments (i.e., *in vitro* fertilization[1] **(IVF)** or gestational surrogacy[2]). There is no known medical treatment to improve the quality of human eggs[3].

To find your next step, please choose the **FIRST** statement that applies to you:

Condition	Action
1. You had a hysterectomy.	• Your fertility assessment is complete; no semen test is needed at this point, since semen examination will be a part of prerequisites for gestational surrogacy.
2. You were told that, for medical reasons, you cannot or should not carry a pregnancy.	You can proceed to gestational surrogacy: **Go to page 95.**
3. You are 42 years or older.	• You are a candidate for *in vitro* fertilization. No semen test is needed at this point, since semen examination will be a part of prerequisites for *in vitro* fertilization. Your next step is an assessment of your endometrial cavity: **Go to page 74.**
4. The above statements do not apply to you.	• Continue with the paragraph below.

Since egg quality can vary from one menstrual cycle to another, optionally, you can repeat the FSH and estradiol tests on your subsequent menstrual periods. You do not need to repeat the ultrasound assessment of your ovaries. With the new result, follow the directions on page 36 and interpretation on pages 38 through 42.

If your test result again shows that you need advanced reproductive treatments (including oocyte donation[4]), you can repeat the test one more time (total of three tests). If your third test result is again similar, you should consider the *in vitro* fertilization treatment.

- Continue on the next page -

From page [ ]   Go to page [ ]

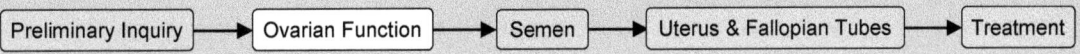

No semen test is required at this point and your next step is an assessment of your endometrial cavity to make sure it can support a pregnancy.

Also, if no sperm are being ejaculated because of the male partner's previous vasectomy, a sperm aspiration[5] procedure will be added to your *in vitro* fertilization treatment.

For instructions on an endometrial cavity assessment: **Go to page 74.**

---

[1] See page 89 for a description of *in vitro* fertilization.
[2] See page 95 for a description of gestational surrogacy.
[3] Clinical evidence suggests that one's lifestyle may influence the quality of eggs being produced. See page 15 for lifestyle recommendations and try to adhere to them as closely as possible.
[4] See page 92 for a description of oocyte (egg) donation.
[5] See page 120 for a description of the sperm aspiration procedure.

Go to page

FERTILITY ASSESSMENT ALGORITHM

Preliminary Inquiry → Ovarian Function → Semen → Uterus & Fallopian Tubes → Treatment

YOU NEED TO CONSIDER USING DONOR EGGS

Your ovarian reserve test results suggest that it is unlikely (but not totally impossible) that you could successfully (pregnancy not ending in a miscarriage) conceive with your own eggs. There is no known medical treatment to improve the quality of human eggs[1]. You need to consider using donor eggs (oocyte donation)[2] to have a live birth.

Since egg quality can vary from one menstrual cycle to another, optionally, you can repeat the FSH and estradiol tests on your subsequent menstrual periods. You do not need to repeat the ultrasound assessment of your ovaries. With the new result, follow the directions on page 36 and interpretation on pages 38 through 42.

If your test result again shows the need to use donor eggs, you can repeat the test one more time (total of three tests). If your third test result is again similar, you should consider the oocyte donation treatment.

To find the next step of your fertility investigation, please choose the **FIRST** statement that applies to you:

Condition	Action
1. You had a hysterectomy.	• Your fertility assessment is complete, you need to consider oocyte donation with gestational surrogacy treatment.
2. You were told that, for medical reasons, you cannot or should not carry a pregnancy.	No semen test is needed at this point, since semen examination will be a part of prerequisites for oocyte donation with gestational surrogacy. You can proceed to oocyte donation with gestational surrogacy treatment: **Go to page 98.**
3. The above statements do not apply to you.	• Your next step is an assessment of your endometrial cavity. No semen test is needed at this point, since semen examination will be a part of prerequisites for oocyte donation. Also, if no sperm are being ejaculated because of previous vasectomy, a sperm aspiration[3] procedure will be added to your oocyte donation treatment. For the assessment of your endometrial cavity: **Go to page 77.**

[1] Clinical evidence suggests that one's lifestyle may influence the quality of eggs being produced. See page 15 for lifestyle recommendations and try to adhere to them as closely as possible.
[2] See page 92 for a description of oocyte (egg) donation.
[3] See page 120 for a description of the sperm aspiration procedure.

From page ☐   Go to page ☐

FERTILITY ASSESSMENT ALGORITHM

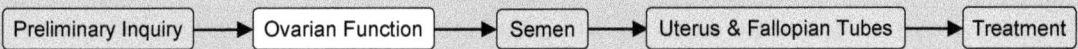

ANOVULATION

If the onsets of your menstrual periods are 40 or more days apart (or you do not have menstrual periods at all), you likely do not produce eggs (anovulation) or do not produce eggs regularly.

Your thyroid gland and the pituitary gland function needs to be assessed together with your glucose-insulin metabolism. You should have the following hormone levels tested:

Test	Description	Result
• TSH	• Thyroid stimulating hormone is produced by the pituitary gland. TSH level increases when thyroid hormone production is low.	
• PRL	• Prolactin hormone is produced by the pituitary gland. Elevated levels of prolactin can cause a lack of ovulation.	
• *Fasting* blood glucose	• Blood sugar level	
• Insulin	• Metabolic hormone produced by the pancreas. Insulin plays a role in the process of ovulation.	

The blood for these tests can be drawn any day of your cycle, first thing in the morning, after fasting overnight.

Once you have obtained the results (typically 2 to 5 days), enter them in the above table and start with interpretation of PRL and TSH levels on the next page:

From page ☐

FERTILITY ASSESSMENT ALGORITHM

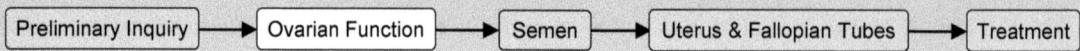

In the table below, select the **FIRST** test result that applies to you:

Condition	Action
1. PRL = 40 or higher (Your prolactin production is too high = hyperprolactinemia) **AND** TSH = 4.2 or higher (Your TSH production is too high. This means that the thyroid hormone production is too low = hypothyroidism)	• Repeat PRL level only. Make sure your blood is drawn early in the morning after fasting overnight. If the repeated PRL is less than 40, consult your physician and only correct your hypothyroidism. If the repeated PRL is again 40 or higher, consult your physician and correct both the hyperprolactinemia and hypothyroidism. If your menstrual periods become regular[1] with the treatment, you started to ovulate and your pregnancy "exposure" starts: **Go to page 34.** • If you still do **not** have regular[1] periods after the hyperprolactinemia and hypothyroidism have been corrected, your pituitary gland FSH production should be interpreted next: **Go to page 49.**
2. PRL = 40 or higher (Your prolactin production is too high = hyperprolactinemia)	• Repeat PRL level. Make sure your blood is drawn early in the morning after fasting overnight. If the repeated PRL is less than 40, your pituitary gland FSH production should be interpreted next: **Go to page 49.** • If the repeated PRL is again 40 or higher, consult your physician and correct the hyperprolactinemia. If your menstrual periods become regular[1] with the treatment, you started to ovulate and your pregnancy "exposure" starts: **Go to page 34.** • If you still do **not** to have regular[1] periods after your hyperprolactinemia has been corrected, your pituitary gland FSH production should be interpreted next: **Go to page 49.**
3. TSH = 4.2 or higher (Your thyroid hormone production is too low = hypothyroidism)	• Consult your physician and correct the hypothyroidism. If your menstrual periods become regular[1] with the treatment, you started to ovulate and your pregnancy "exposure" starts: **Go to page 34.** • If you still do **not** to have regular[1] periods after your hypothyroidism has been corrected, your pituitary gland FSH production should be interpreted next: **Go to page 49.**
4. None of the above statements apply to you.	• Your pituitary gland FSH production should be interpreted next: **Go to page 49.**

[1] Variation of no more than seven days between the shortest and the longest cycle and less than 40 days apart.

Go to page

FERTILITY ASSESSMENT ALGORITHM

Preliminary Inquiry → Ovarian Function → Semen → Uterus & Fallopian Tubes → Treatment

CHECKING FOR LOW FSH AND ESTROGEN PRODUCTION

Normal production of the follicle stimulating hormone (FSH) by the pituitary gland is necessary for regular maturation and release of eggs from the ovaries. Insufficient FSH production can be caused by low body fat content, excessive physical activity, stress, and other factors.

Normal estradiol (estrogen, E_2) production is related to the FSH production and is necessary for normal development of the endometrial lining.

First, copy your FSH and estradiol levels from "Hormonal test #1" entry on page 37 here:

Test	Result
• FSH	
• Estradiol (estrogen, E_2)	

Use the following table to interpret your results:

Condition	Action
• FSH = less than 5.0 **AND** • E_2 = less than 25 (FSH and estradiol production is too low = hypothalamic hypogonadism)	• If you are underweight, bring your weight up. If you exercise very strenuously, reduce your physical activity. If your menstrual periods become regular[1], you started to ovulate and your pregnancy "exposure" starts: **Go to page 34.** • If you still do **not** to have regular[1] menstrual periods: **Go to page 50.**
• FSH = 5.0 or higher **OR** • E_2 = 25 or higher	• Your lack of regular[1] menstrual periods (ovulations) is not due to low FSH or estradiol production: **Go to page 51.**

[1] Variation of no more than seven days between the shortest and the longest cycle and less than 40 days apart.

From page ☐ Go to page ☐

49

HYPOTHALAMIC HYPOGONADISM

Hypothalamic hypogonadism is a condition in which part of the brain (hypothalamus) which stimulates the pituitary gland (a pea-sized gland at the base of the brain) sends mixed signals to the pituitary.

This irregular stimulation of the pituitary gland results in an inadequate production of FSH (follicle stimulating hormone) and LH (luteinizing hormone) by the pituitary. Without normal FSH and LH secretion the ovaries do not produce eggs and normal levels of estrogen and progesterone hormones. The ovarian function is suboptimal (hypogonadism).

Low body weight, strenuous exercise, and high levels of stress are the most common reasons for hypothalamic hypogonadism. Sometimes, its cause is unexplained.

Fortunately, most hypothalamic hypogonadism patients who plan to conceive will respond to a treatment that supplies the FSH and LH hormones (ovarian stimulation with injectable medications). To determine whether you are a candidate for this treatment, the male partner's semen quality must be assessed next: **Go to page 53.**

FERTILITY ASSESSMENT ALGORITHM

Preliminary Inquiry → Ovarian Function → Semen → Uterus & Fallopian Tubes → Treatment

POLYCYSTIC OVARIES

Since you have ruled out hypothyroidism, hyperprolactinemia, and hypothalamic hypogonadism as the cause of your anovulation, it is very likely that your lack of ovulation is caused by polycystic ovaries (PCO). Polycystic ovaries are *by far* the most common cause of anovulation. The term "polycystic" refers to the increased number of ovarian follicles (not cysts) present within the ovaries of most women with PCO.

Anecdotal evidence suggests that perhaps 5% to 15% of all women are born with some degree of polycystic ovaries. Many, if not most of these women, will never find out since they will be able to conceive. These women have a mild degree of PCO.

Women with more severe PCO will, at least temporarily, lose the regularity of their menstruations (become anovulatory) and may need medical help to conceive. In its most severe forms, PCO could make it quite difficult for a woman to ovulate and conceive even with ovarian stimulation.

The lack of ovulation is accompanied by a hormonal imbalance which, in some women, involves abnormal insulin production and glucose metabolism.

To determine whether your insulin and glucose production is abnormal and whether you may benefit from improving it by taking Metformin (Glucophage, an oral anti-diabetic drug), divide the fasting glucose level from your test results on page 47 by the blood level of insulin:

$$\text{Glucose to insulin ratio} = \frac{\text{Fasting glucose}}{\text{Insulin}}$$

Using the calculated ratio, find your next step in the table:

Glucose to insulin ratio	Action
• 4.5 or lower = insulin resistance (You could benefit from Metformin treatment)	• Consult your physician and start taking Metformin. If your menstrual periods become regular[1] on Metformin, you are ovulating and your pregnancy "exposure" starts: **Go to page 34.** • Even if you still do **not** have regular[1] periods, stay on Metformin as directed by your physician. To determine which of the available treatments is most appropriate for you, the male partner's semen quality must be assessed next: **Go to page 53.**
• Greater than 4.5 (Likely no benefit from Metformin treatment)	• To determine which of the available treatments is most appropriate for you, the male partner's semen quality must be assessed next: **Go to page 53.**

[1] Variation of no more than seven days between the shortest and the longest cycle and less than 40 days apart.

From page ☐   Go to page ☐

FERTILITY ASSESSMENT ALGORITHM

Preliminary Inquiry → Ovarian Function → Semen → Uterus & Fallopian Tubes → Treatment

NO SPERM ARE BEING EJACULATED BECAUSE OF PREVIOUS VASECTOMY

At this point, you will need to decide whether you should have a sperm aspiration[1] procedure or microsurgical reversal of the vasectomy. You should discuss the pros and cons of either treatment with your urologist.

Sperm can be aspirated from the epididymis (convoluted tube, part of the spermatic duct system) or testes. The sperm aspiration procedure is combined with one of the advanced reproductive treatments (i.e., *in vitro* fertilization[2], IVF). It is normally an uncomplicated, quick outpatient procedure done by a urologist and requires only a small amount of local anesthetic.

Alternatively, you may choose a microsurgical repair of the vasectomy which can remove blockage in the sperm delivery system. It is a complex procedure done by a urologist specializing in male infertility treatments.

If you decide to have the sperm aspiration procedure, you will need *in vitro* fertilization and your next step is an assessment of female partner's endometrial cavity: **Go to page 74.**

If you decide to have the vasectomy surgically repaired and *motile sperm are produced* after the surgery, find your next step in the table:

Condition	Action
• The typical number of days between onsets of your periods is 40 or more days. • You do not have menstrual periods.	• Your next step is the investigation of the lack of regular[3] ovulations: **Go to page 47.**
• The above statements do not apply to you.	**Go to page 35.**

[1] See page 120 for a description of the sperm aspiration procedure.
[2] See page 89 for a description of *in vitro* fertilization.
[3] Variation of no more than seven days between the shortest and the longest cycle and menstrual periods less than 40 days apart.

From page ☐   Go to page ☐

FERTILITY ASSESSMENT ALGORITHM

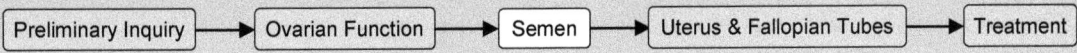

SEMEN ANALYSIS

The result of your ovarian reserve testing suggests that you should be able to conceive with your eggs.

Since a significant portion of all infertility is due to male factor infertility and since the presence of male infertility has a major impact on the direction of fertility testing and treatment, a semen assessment should be your next test.

Due to a sometimes profound variation in the semen parameters, two or more semen analyses should be done at least three days apart. The testing laboratory will typically ask for a three day abstinence from ejaculation prior to semen collection. The specimen must be collected by masturbation.

Once you have obtained the semen analysis result, enter the values for semen volume, sperm concentration (sperm count), and the percentage of total motility (percentage of all moving sperm) in the table below.

If you had one or more semen analyses within the last two years, you can use those results. If more than two semen analyses results are available, use the values for the two most recent ones.

	First Semen Analysis	Second Semen Analysis	Units
• Date			
• Volume			ml
• Concentration (sperm count)			million/ml
• Total motility			%
• Total motile sperm			Millions per ejaculate

Calculate the **total motile sperm** per ejaculate by multiplying the semen volume in milliliters (ml) times sperm concentration (sperm count) in millions per ml times fraction of total motile sperm.

- Continue on the next page -

From page ☐

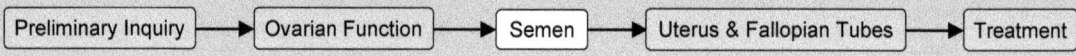

For example: Volume = 3.2ml

Sperm concentration = 30 million/ml

Total motility = 30% (= 0.3 fraction)

3.2 x 30 x 0.3 = 28.8 (total motile sperm in millions)

In the following table, use the **HIGHER** of the two total motile sperm values:

Total motile sperm per ejaculate	Action
• 4 million or more	Go to page 55
• More than 0 and less than 4 million	Go to page 56
• No sperm or no motile sperm found	Go to page 57

Go to page ☐

FERTILITY ASSESSMENT ALGORITHM

Preliminary Inquiry → Ovarian Function → Semen → Uterus & Fallopian Tubes → Treatment

FOUR OR MORE MILLION MOTILE SPERM PER EJACULATE

The semen test result indicates a *probable* absence of male infertility. It is important to note that semen analysis cannot *absolutely* rule out male factor infertility.

Use the table below to find the next step in your fertility investigation:

Condition	Action
• You can conceive *without difficulty* and you have had *three or more* miscarriages.	• The next step in your fertility investigation is an assessment of your uterine (endometrial) cavity: **Go to page 79.**
• The above statement does not apply to you.	• The next step in your fertility investigation is an assessment of your uterus and Fallopian tubes with a hysterosalpingogram (HSG, x-ray of the uterus and Fallopian tubes): **Go to page 60.**

From page [ ]   55   Go to page [ ]

FERTILITY ASSESSMENT ALGORITHM

Preliminary Inquiry → Ovarian Function → Semen → Uterus & Fallopian Tubes → Treatment

MORE THAN ZERO AND LESS THAN FOUR MILLION MOTILE SPERM PER EJACULATE

The semen test result indicates a *probable* presence of male infertility. You should consider *in vitro* fertilization[1] (IVF) as the most appropriate treatment.

Typically, the male fertility *cannot* be increased by taking medications or surgically (i.e. removal of a varicocele) but it is now possible to treat all but the most severe male factors with *in vitro* fertilization and intracytoplasmic sperm injection[2] (ICSI).

Before the IVF treatment, the uterine cavity must be checked for presence of endometrial polyps, internal myomas (fibroids, hard nodules inside the uterine cavity), and intrauterine adhesions (scarring): **Go to page 74.**

---

[1] See page 89 for a description of *in vitro* fertilization.
[2] ICSI is a micromanipulation technique in which a single sperm is inserted directly into an egg.

From page [ ] Go to page [ ]

FERTILITY ASSESSMENT ALGORITHM

Preliminary Inquiry → Ovarian Function → Semen → Uterus & Fallopian Tubes → Treatment

NO MOTILE SPERM IN SEMEN ANALYSIS

Either there were no sperm seen (azoospermia) in the semen samples or sperm were present but were not moving (non-motile sperm). The male partner's further testing involves the following two steps:

1. Another semen sample **and** a post-ejaculation urine specimen (to rule out retrograde ejaculation) should be collected and be centrifuged by the laboratory prior to examination. The concentrated "pellets" at the bottom of the centrifuge tubes from these two samples are then examined for signs of motile sperm:

Findings	Action
• Any number of motile sperm seen in either pellet.	• Do **not** go to step #2 below: **Go to page 56.**
• No motile sperm seen.	• Continue with step #2 below.

2. The male partner needs to see a urologist. He will order tests to differentiate between the inability of the testes to produce sperm and blockage in the sperm delivery system. He will also order tests to rule out chromosomal (genetic) abnormalities of the male partner:

Findings	Action
• Only non-motile sperm are produced.	• The male partner will need a sperm aspiration procedure[1]: **Go to page 58.**
• There is a blockage in the sperm delivery system.	• Sperm aspiration[1] from above the blockage: **Go to page 58.** OR • Microsurgical repair by your urologist[2]. Once motile sperm are produced after the surgery: **Go to page 35.**
• The testes cannot produce sperm.	• No treatment is available and you will need to use donor semen in your treatment. The donor semen is typically obtained from a sperm bank: **Go to page 59.**

[1] See page 120 for a description of the sperm aspiration procedure.
[2] Microsurgical repair can remove blockage in the sperm delivery system. It is a complex procedure done by a urologist specializing in male infertility treatments.

From page ☐ Go to page ☐

FERTILITY ASSESSMENT ALGORITHM

Preliminary Inquiry → Ovarian Function → Semen → Uterus & Fallopian Tubes → Treatment

SPERM ASPIRATION

The semen test result indicates significant male factor infertility. You will need a sperm aspiration procedure[1] which must be combined with one of the advanced reproductive treatments (i.e., *in vitro* fertilization[2], IVF).

Sperm aspiration is normally an uncomplicated, quick outpatient procedure requiring only a small amount of local anesthetic. It is done by a urologist.

Prior to the IVF treatment, your uterine cavity will need to be checked for the presence of endometrial polyps, internal myomas (fibroids, hard nodules inside the uterine cavity), and intrauterine adhesions (scarring): **Go to page 74.**

---

[1] See page 120 for a description of the sperm aspiration procedure.
[2] See page 89 for a description of *in vitro* fertilization.

From page ☐   Go to page ☐

NO SPERM PRODUCTION IS POSSIBLE

You and your partner will need to use donor semen to be able to conceive. Typically, donor semen is obtained from a sperm bank, but you could also find a sperm donor on your own.

The Federal Drug Administration (FDA) has set up strict testing requirements for sperm donors. If you decide not to use a sperm bank, your physician will advise you on the most current sperm donor screening requirements.

Using donor semen can be as simple as cervical or intrauterine insemination timed to your ovulations. You do not need an infertility specialist for donor semen inseminations, they can be done by your gynecologist.

The function of your uterus and Fallopian tubes must be normal to conceive with donor semen insemination. Your next step should be their assessment with a hysterosalpingogram (HSG, x-ray of the uterus and Fallopian tubes): **Go to page 60.**

FERTILITY ASSESSMENT ALGORITHM

Preliminary Inquiry → Ovarian Function → Semen → Uterus & Fallopian Tubes → Treatment

HYSTEROSALPINGOGRAM (HSG)

Hysterosalpingogram (HSG, x-ray of the uterus and Fallopian tubes) is a test to evaluate the inside of your uterus (endometrial cavity) and the Fallopian tubes (the connection between the ovaries and uterus). Your endometrial cavity must be free of polyps, myomas (fibroids, hard nodules inside the uterine cavity), and intrauterine adhesions (scarring). In addition, the Fallopian tubes must be open to allow for meeting of the egg and sperm.

HSG is a quick procedure done in a radiology department. If you have had an HSG within the last 12 months, you do not need to repeat it.

During the 10 to 20 minute HSG test, a thin tube is passed through the cervical canal inside the uterine cavity and radio-opaque dye is slowly instilled. Two or more x-ray pictures are taken. You may experience menstrual-like cramping when the tube is being passed and when the dye is instilled. The result can be interpreted right away.

Ask the radiologist (and your physician if s/he will be reading the HSG images as well) to focus on the isthmic portion of your Fallopian tubes (the segment of tubes adjacent to the uterus) to look for signs of "extra channels", filling defects, and out-pouching of the tubal lumen. If any of these are found in the isthmic portion of either tube, you have salpingitis isthmica nodosa. This is a rare condition whose causes are poorly understood and the HSG images must be **carefully evaluated** in order to detect it.

1. If the HSG findings were normal, your fertility assessment is complete: **Go to page 73.**

2. If the HSG showed only **ONE** of the following conditions:

Findings	Description	Action
• Filling defect	Polyp(s), fibroid(s) or scaring in the endometrial cavity.	Go to page 62
• Pelvic adhesions	Scaring outside the uterus.	Go to page 63
• Proximal tubal block	One or both Fallopian tubes did not fill with dye.	Go to page 63
• Hydrosalpinx	One or both Fallopian tubes are blocked at the far end (clubbed tubes).	Go to page 65
• Salpingitis isthmica nodosa	See text above.	Go to page 66

- Continue on the next page -

From page [ ]   Go to page [ ]

FERTILITY ASSESSMENT ALGORITHM

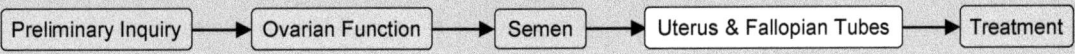

3. If the HSG showed any **COMBINATION** of the above conditions, select the row and column in the table below that best describe your HSG findings:

Findings	Filling defect	Pelvic adhesions	Proximal tubal block	Hydrosalpinx
• Pelvic adhesions	Go to page 67	↓	↓	↓
• Proximal tubal block	Go to page 67	Go to page 63	↓	↓
• Hydrosalpinx	Go to page 69	Go to page 65	Go to page 65	↓
• Salpingitis isthmica nodosa	Go to page 72	Go to page 66	Go to page 66	Go to page 71

Go to page ☐

FILLING DEFECT

A "filling defect" on HSG means that something keeps the x-ray dye from being able to fully fill the uterine cavity. There are three conditions that can result in filling defects: endometrial polyps, internal myomas (fibroids, hard nodules inside the uterine cavity), and intrauterine adhesions (scarring). Regardless of the cause of the filling defect, the underlying structure should be removed otherwise it could reduce the probability of embryo implantation.

On the other hand, the "filling defect" should not be considered as the sole cause of infertility. After its correction, the recommended treatment is the same as if the endometrial cavity had been normal.

Normally, you will be scheduled for a hysteroscopy to remove the cause of the filling defect. Hysteroscopy is a quick outpatient procedure requiring only very light anesthesia. A thin optical scope is passed through the cervical canal inside the uterus and the endometrial cavity is visualized. The cause of the filling defect is identified and removed. The recovery time after this procedure is typically very short.

The hysteroscopy may also find that your uterine cavity is normal and that the HSG's "abnormal" findings were an artifact (error).

Once your uterine cavity is normal, your fertility assessment is complete and you can proceed to the next step: **Go to page 73.**

PELVIC ADHESIONS AND/OR PROXIMAL TUBAL BLOCK

Your hysterosalpingogram showed either signs of pelvic adhesions (pooling of x-ray dye), a lack of fill of your Fallopian tube(s) (proximal tubal block), or both. This could be an artifact (error), a uterine muscle spasm may have prevented the x-ray dye from entering the Fallopian tube(s) and pooling of dye on HSG does not always mean that pelvic adhesions are present. You will need a laparoscopy (see below) to differentiate between an artifact and a true pathology.

Alternatively, you may decide that you do not want to further explore the condition of your Fallopian tubes and/or the possible presence of pelvic adhesions and that you want to proceed directly to *in vitro* fertilization[1] (IVF). IVF bypasses the function of the Fallopian tubes and the presence of tubal blockage or pelvic adhesions is inconsequential in this treatment. Under these circumstances, you would *not* need to have laparoscopy and your fertility assessment would be complete: **Go to page 89.**

If the lack of fill of one of the Fallopian tubes is due to the removal of the tube (whether or not the HSG also showed signs of pelvic adhesion) you should consider *in vitro* fertilization as the treatment most likely to result in a pregnancy. You do not need a laparoscopy and your fertility assessment is complete: **Go to page 89.**

Laparoscopy is an outpatient procedure requiring mild general anesthesia. During the procedure, a thin optical scope is inserted inside the abdomen through a very small incision just below the bellybutton and the pelvic organs are examined. Special dye is used to determine if the Fallopian tubes are open. The recovery time after this procedure is typically very short.

If the laparoscopy finds that one or both Fallopian tubes are blocked at the **distal** (far) end, you have a hydrosalpinx (clubbed tube). If the hydrosalpinx is 10 millimeters or wider, **the Fallopian tube should be removed or "clipped"** next to the uterus during the laparoscopy. This severs the communication between the inside of the Fallopian tube and the inside of the uterus. The secretion produced by the lining of the hydrosalpinx is embryo-toxic and could impede embryo implantation. You should discuss the possibility of clipping or removing your Fallopian tube(s) with your physician *prior* to the laparoscopy.

Once you have reviewed the laparoscopy findings with your physician, in the table on the next page, select the **FIRST** condition that applies to you:

---

[1] See page 89 for a description of *in vitro* fertilization.

FERTILITY ASSESSMENT ALGORITHM

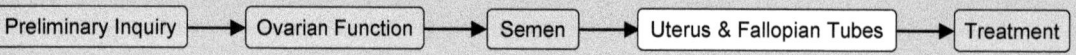

Laparoscopy Findings

Findings	Action
1. One or both Fallopian tubes are blocked at the distal end (clubbed tube, hydrosalpinx).	• Your fertility assessment is complete. You should consider *in vitro* fertilization[1] (IVF) as the treatment most likely to result in a pregnancy: **Go to page 89.**
2. One or both Fallopian tubes did not fill (proximal tubal block).	
3. Pelvic adhesions (scaring)	
4. Endometriosis[2]	• These are typically **not** a cause of infertility and you should not need major surgery (laparotomy[4]) to have them removed. Your fertility assessment is complete. To find your treatment options: **Go to page 73.**
5. External fibroids[3]	
6. Your laparoscopy findings were normal; the HSG findings were an artifact (error).	• Your fertility assessment is complete. To find your treatment options: **Go to page 73.**

[1] See page 89 for a description of *in vitro* fertilization.
[2] The presence of tissue that normally grows inside the uterus in an abnormal anatomical location.
[3] Benign growths of muscle cells that develop on the outside uterine wall.
[4] Major abdominal surgery to correct abnormalities of the reproductive organs.

Go to page ☐

HYDROSALPINX

If the HSG showed that one or both Fallopian tubes are blocked at the distal (far) end, you have a hydrosalpinx (clubbed tube). If one of the Fallopian tubes also did not fill, you have a combination of a proximal block and a hydrosalpinx. Either one means that at some point in the past you had an infection inside your Fallopian tubes (pelvic inflammatory disease). You will need to bypass your Fallopian tubes with *in vitro* fertilization[1] (IVF) treatment. Hydrosalpinx may also co-exist with pelvic adhesions.

Having surgery to repair your fallopian tubes is not recommended. It may be a long time (up to two to five years) before you may see any possible benefit of the surgery (pregnancy) and the success rates of surgical hydrosalpinx repair are typically low.

In most cases it may be more efficacious and, in the long run, less expensive to bypass the Fallopian tubes altogether with *in vitro* fertilization. If the Fallopian tubes are the main cause of infertility, the IVF pregnancy rates are typically quite high.

If the hydrosalpinx is 10 millimeters or wider on the x-ray, **the Fallopian tube should be removed or "clipped"** next to the uterus before the IVF treatment. This severs the communication between the inside of the Fallopian tube and the inside of the uterus. The secretion produced by the inner lining of the hydrosalpinx is embryo-toxic and could impede embryo implantation.

You will need a laparoscopy to have the Fallopian tube(s) removed or clipped. Laparoscopy is an outpatient procedure requiring mild general anesthesia. During the procedure, a thin optical scope is inserted inside the abdomen through a very small incision just below the bellybutton. Specially designed instruments are used to remove or clip the Fallopian tube(s). The recovery time after this procedure is typically very short.

Once the tube(s) has/have been clipped or removed (if needed), your fertility assessment is complete and you should proceed to *in vitro* fertilization treatment: **Go to page 89.**

---

[1] See page 89 for a description of *in vitro* fertilization.

FERTILITY ASSESSMENT ALGORITHM

Preliminary Inquiry → Ovarian Function → Semen → Uterus & Fallopian Tubes → Treatment

SALPINGITIS ISTHMICA NODOSA

Salpingitis isthmica nodosa is a slowly progressing permanent swelling of the isthmic (adjacent to the uterus) portion of the Fallopian tubes. Its cause is unknown but it could represent a residual effect of a pelvic inflammatory disease (some of which may occur without any symptoms). Salpingitis isthmica nodosa may also co-exist with pelvic adhesions.

After many years, the swelling may sever the communication between the inside of the Fallopian tubes and the uterine cavity (proximal block). That Fallopian tube will not fill on the HSG. Even if this communication still exists, it is VERY rare for the tubes to function normally. Their function must be bypassed with *in vitro* fertilization[1] (IVF) treatment.

Your fertility assessment is complete and you should proceed to *in vitro* fertilization treatment: **Go to page 89.**

---

[1] See page 89 for a description of *in vitro* fertilization.

From page [ ] Go to page [ ]

FERTILITY ASSESSMENT ALGORITHM

Preliminary Inquiry → Ovarian Function → Semen → **Uterus & Fallopian Tubes** → Treatment

FILLING DEFECT AND PELVIC ADHESIONS
OR
FILLING DEFECT AND PROXIMAL TUBAL BLOCK

A "filling defect" on HSG means that something keeps the x-ray dye from being able to fully fill the uterine cavity. There are three conditions that can result in filling defects: endometrial polyps, internal myomas (fibroids, hard nodules inside the uterine cavity), and intrauterine adhesions (scarring). Regardless of the cause of the filling defect, the underlying structure should be removed otherwise it could reduce the probability of embryo implantation.

On the other hand, the "filling defect" should not be considered as the sole cause of infertility. After its correction, the recommended treatment is the same as if the endometrial cavity had been normal.

Your HSG also showed signs of pelvic adhesions (pooling of x-ray dye) and/or a lack of fill of your Fallopian tube(s) (proximal tubal block). This could be an artifact (error), a uterine muscle spasm may have prevented the x-ray dye from entering the Fallopian tube(s) and pooling of dye on HSG does not always mean that pelvic adhesions are present. You will need a laparoscopy (see below) to differentiate between an artifact and a true pathology.

Alternatively, you may decide that you do not want to further explore the condition of your Fallopian tubes and/or the possible pelvic adhesions and that you want to proceed directly to *in vitro* fertilization[1] (IVF). IVF bypasses the function of the Fallopian tubes and the presence of tubal blockage or pelvic adhesions is inconsequential for this treatment. Under these circumstances, you would *not* need to have laparoscopy; however you will need a hysteroscopy to remove the cause of the filling defect: **Go to page 76.**

If the lack of fill of one of the Fallopian tubes is due to a previous removal of the tube (whether or not the HSG also showed signs of pelvic adhesion) you should consider *in vitro* fertilization as the treatment most likely to result in a pregnancy. You do not need a laparoscopy and you only need a hysteroscopy to remove the cause of the filling defect: **Go to page 76.**

Hysteroscopy and a laparoscopy can be done together as an outpatient procedure requiring mild general anesthesia. In the hysteroscopy portion of the procedure, a thin optical scope is passed through the cervical canal inside the uterus and the endometrial cavity is visualized. The cause of the filling defect is identified and removed.

- Continue on the next page -

FERTILITY ASSESSMENT ALGORITHM

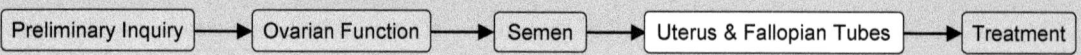

The hysteroscopy may also find that your uterine cavity is normal and that the HSG's "abnormal" findings were an artifact.

During the laparoscopy portion, a thin optical scope is inserted inside the abdomen through a very small incision just below the bellybutton and the pelvic organs are examined. Special dye is used to determine if the Fallopian tubes are open. The recovery time after this procedure is typically very short.

If the laparoscopy shows that one or both Fallopian tubes are blocked at the distal (far) end, you have a hydrosalpinx (clubbed tube). If the hydrosalpinx is 10 millimeters or wider, **the Fallopian tube should be removed or "clipped"** next to the uterus at this time. This severs the communication between the inside of the Fallopian tube and the inside of the uterus. The secretion produced by the inner lining of the hydrosalpinx is embryo-toxic and could impede embryo implantation. You should discuss the possibility of clipping or removing your Fallopian tube(s) with your physician *prior* to the laparoscopy.

Once your uterine cavity is normal and you have reviewed the laparoscopy findings with your physician, in the table below, select the **FIRST** condition that applies to you:

Findings	Action
1. One or both Fallopian tubes are blocked at the distal end (clubbed tube, hydrosalpinx).	• Your fertility assessment is complete. You should consider *in vitro* fertilization[1] (IVF) as the treatment most likely to result in a pregnancy: **Go to page 89.**
2. One or both Fallopian tubes did not fill (proximal tubal block).	
3. Pelvic adhesions (scaring)	
4. Endometriosis[2]	• These are typically **not** a cause of infertility and you should not need major surgery (laparotomy[4]) to have them removed. Your fertility assessment is complete. To find your treatment options: **Go to page 73.**
5. External fibroids[3]	
6. Your laparoscopy findings were normal; the HSG findings were an artifact (error).	• Your fertility assessment is complete. To find your treatment options: **Go to page 73.**

[1] See page 89 for a description of *in vitro* fertilization.
[2] The presence of tissue that normally grows inside the uterus in an abnormal anatomical location.
[3] Benign growths of muscle cells that develop on the outside uterine wall.
[4] Major abdominal surgery to correct abnormalities of the reproductive organs.

Go to page

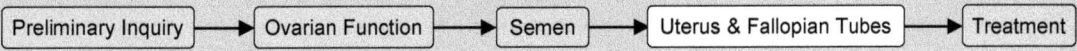

FILLING DEFECT AND HYDROSALPINX

A "filling defect" on HSG means that something keeps the x-ray dye from being able to fully fill the uterine cavity. There are three conditions that can result in filling defects: endometrial polyps, internal myomas (fibroids, hard nodules inside the uterine cavity), and intrauterine adhesions (scarring). Regardless of the cause of the filling defect, the underlying structure should be removed otherwise it could reduce the probability of embryo implantation.

On the other hand, the "filling defect" should not be considered as the sole cause of infertility. After its correction, the recommended treatment is the same as if the endometrial cavity had been normal.

Your hysterosalpingogram also showed that one or both Fallopian tubes are blocked at the distal (far) end (hydrosalpinx, clubbed tube). If one of the Fallopian tubes also did not fill, you have a combination of a proximal block and hydrosalpinx. Either one means that at some point in the past you had an infection inside your Fallopian tubes (pelvic inflammatory disease). You will need to bypass your Fallopian tubes with *in vitro* fertilization[1] (IVF) treatment.

Having surgery to repair your fallopian tubes is not recommended. It may be a long time (up to two to five years) before you may see any possible benefit of the surgery (pregnancy) and the success rates of surgical hydrosalpinx repair are typically low.

In most cases it may be more efficacious and, in the long run, less expensive to bypass the Fallopian tubes altogether with *in vitro* fertilization. If the Fallopian tubes are the main cause of infertility, the IVF pregnancy rates are typically quite high.

If the hydrosalpinx is 10 millimeters or wider on the x-ray, **the Fallopian tube should be removed or "clipped"** next to the uterus before the IVF treatment. This severs the communication between the inside of the Fallopian tube and the inside of the uterus. The secretion produced by the inner lining of the hydrosalpinx is embryo-toxic and could impede embryo implantation.

If your hydrosalpinx is less than 10 millimeters, it may not need to be removed or clipped and your doctor may only schedule a hysteroscopy to remove the cause of the filling defect.

Hysteroscopy is a quick outpatient procedure requiring only very light anesthesia. A thin optical scope is passed through the cervical canal inside the uterus and the endometrial cavity is visualized.

- Continue on the next page -

From page ☐

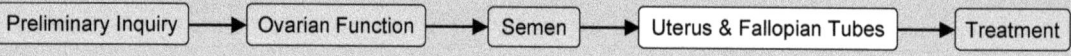

The cause of the filling defect is identified and removed. The hysteroscopy may also find that your uterine cavity is normal and that the HSG's "abnormal" findings were an artifact (error). The recovery time after this procedure is typically very short.

If the hydrosalpinx is 10 millimeters or wider on the x-ray, your doctor may schedule a combination of a hysteroscopy and a laparoscopy. This is also an outpatient procedure requiring only mild general anesthesia.

During the laparoscopy portion, a thin optical scope is inserted inside the abdomen through a very small incision just below the bellybutton and the pelvic organs are examined. Specially designed instruments are used to remove or "clip" the Fallopian tube(s). The recovery time after this procedure is typically very short.

Once your uterine cavity is normal and your Fallopian tube(s) has/have been clipped or removed (if needed), your fertility assessment is complete and you can proceed to *in vitro* fertilization treatment: **Go to page 89.**

---

[1] See page 89 for a description of *in vitro* fertilization.

Go to page

HYDROSALPINX AND SALPINGITIS ISTHMICA NODOSA

If the HSG showed that one or both Fallopian tubes are blocked at the distal (far) end, you have a hydrosalpinx (clubbed tube). Your HSG also showed signs of salpingitis isthmica nodosa. This is a slowly progressing permanent swelling of the isthmic (adjacent to the uterus) portion of the Fallopian tubes. Its cause is unknown but it could represent a residual effect of a pelvic inflammatory disease (some of which may occur without any symptoms).

After many years, the swelling may sever the communication between the inside of the Fallopian tubes and the uterine cavity (proximal block). That Fallopian tube will not fill on the HSG. Even if this communication still exists, it is VERY rare for the tubes to function normally. Their function must be bypassed with *in vitro* fertilization[1] (IVF) treatment. If the Fallopian tubes are the main cause of infertility, the IVF pregnancy rates are typically quite high.

If the hydrosalpinx is 10 millimeters or wider on the x-ray, **the Fallopian tube should be removed or "clipped"** next to the uterus before the IVF treatment. This severs the communication between the inside of the Fallopian tube and the inside of the uterus. The secretion produced by the inner lining of the hydrosalpinx is embryo-toxic and could impede embryo implantation.

You will need a laparoscopy to have the Fallopian tube(s) removed or clipped. Laparoscopy is an outpatient procedure requiring mild general anesthesia. During the procedure, a thin optical scope is inserted inside the abdomen through a very small incision just below the bellybutton and the pelvic organs are examined. Specially designed instruments are used to remove or clip the Fallopian tube(s).

Once the tube(s) has/have been clipped or removed (if needed), your fertility assessment is complete and you should proceed to *in vitro* fertilization treatment: **Go to page 89.**

---

[1] See page 89 for a description of *in vitro* fertilization.

FILLING DEFECT AND SALPINGITIS ISTHMICA NODOSA

A "filling defect" on HSG means that something keeps the x-ray dye from being able to fully fill the uterine cavity. There are three conditions that can result in filling defects: endometrial polyps, internal myomas (fibroids, hard nodules inside the uterine cavity), and intrauterine adhesions (scarring). Regardless of the cause of the filling defect, the underlying structure should be removed otherwise it could reduce the probability of embryo implantation.

On the other hand, the "filling defect" should not be considered as the sole cause of infertility. After its correction, the recommended treatment is the same as if the endometrial cavity had been normal.

Normally, you will be scheduled for a hysteroscopy to remove the cause of the filling defect. Hysteroscopy is a quick outpatient procedure requiring only very light anesthesia. A thin optical scope is passed through the cervical canal inside the uterus and the endometrial cavity is visualized. The cause of the filling defect is identified and removed. The recovery time after this procedure is typically very short.

The hysteroscopy may also find that your uterine cavity is normal and that the HSG's "abnormal" findings were an artifact (error).

Salpingitis isthmica nodosa is a slowly progressing permanent swelling of the isthmic (adjacent to the uterus) portion of Fallopian tubes. Its cause is unknown but it could represent a residual effect of a pelvic inflammatory disease (some of which may occur without any symptoms).

After many years, the swelling may sever the communication between the inside of the Fallopian tubes and the uterine cavity. Even if this communication still exists, it is VERY rare for the tubes to perform normally. Their function must be bypassed with *in vitro* fertilization[1] (IVF) treatment.

Once your uterine cavity is normal, your fertility assessment is complete and you can proceed to *in vitro* fertilization: **Go to page 89.**

---

[1] See page 89 for a description of *in vitro* fertilization.

FERTILITY ASSESSMENT FORMULA

Preliminary Inquiry → Ovarian Function → Semen → Uterus & Fallopian Tubes → Treatment

YOU HAVE NORMAL ENDOMETRIAL CAVITY
AND
YOUR FALLOPIAN TUBES ARE PATENT

The results of your hysterosalpingogram or laparoscopy were either normal or you had a "filling defect" which has been corrected or your laparoscopy found only endometriosis[1] or external myomas (fibroids)[2].

Your uterus should be able to support embryo implantation and your Fallopian tubes should allow meeting of the eggs and sperm.

To find your next step, select the **FIRST** statement that best describes your condition:

Condition	Action
1. Your FSH and estrogen production is too low, you have hypothalamic hypogonadism[3] (diagnosis from page 50).	• Your fertility assessment is complete. You can proceed to treatment for hypothalamic hypogonadism: **Go to page 84.**
2. You need donor semen to conceive (diagnosis from page 59).	• Your fertility assessment is complete. You can proceed to ovarian stimulation with oral medications: **Go to page 85.**
3. The typical number of days between onsets of your periods is 40 or more days.	
4. None of the above statements apply to you.	• Your fertility assessment is complete. You can proceed to the treatment options page: **Go to page 82.**

[1] The presence of tissue that normally grows inside the uterus in an abnormal anatomical location.
[2] Benign growths of muscle cells that develop on the outside uterine wall.
[3] In hypothalamic hypogonadism, the pituitary gland does not produce adequate amounts of follicle stimulating hormone (FSH) and luteinizing hormone (LH) resulting in a lack of ovulation.

From page ☐   Go to page ☐

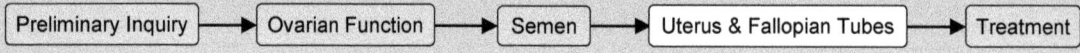

ENDOMETRIAL CAVITY

It is important to confirm the absence of polyps[1], scarring, and fibroids (myomas)[2] *inside* the uterus. The presence of any of these structures could reduce the probability of embryo implantation. On the other hand, they should not be considered as the sole cause of infertility. After their removal, the recommended treatment is the same as if the endometrial cavity had been normal.

Sonohysterogram, hysterosalpingogram (see below), or hysteroscopy[3] can be used to assess the endometrial cavity. Only *one* of these tests is necessary. If you have had *any* of them within the last six months, you do not need another one; otherwise you should have either a sonohysterogram or hysterosalpingogram.

- Sonohysterogram (a transvaginal ultrasound of the uterus) is the least invasive and least expensive of the above tests. It is typically done by your gynecologist. It can be done on any cycle day when you are not having your period. The ideal time for this test is just after the end of a menstrual period.

  A thin, soft catheter is passed through the cervix inside the uterus and a small amount of sterile saline solution is instilled. This allows the contours of the endometrial cavity to be visualized.

  If the ultrasound shows that one or both Fallopian tubes are distended to 10 millimeters or wider, **the Fallopian tube(s) should be removed or "clipped"** next to the uterus before the IVF treatment. This severs the communication between the inside of the Fallopian tube(s) and the inside of the uterus. The secretion produced by the inner lining of the distended Fallopian tubes is embryo-toxic and could impede embryo implantation. You will need a laparoscopy[4] to have the Fallopian tube(s) removed or clipped.

- Hysterosalpingogram (HSG) is an x-ray of the uterus and Fallopian tubes. It is a quick procedure done in a radiology department.

  During the 10 to 20 minute test, a thin tube is passed through the cervical canal inside the uterine (endometrial) cavity and radio-opaque dye is slowly instilled. Two or more x-ray pictures are taken. You may experience menstrual-like cramping when the tube is being passed and when the dye is instilled. The result is available immediately and can be interpreted right away.

  If the HSG showed that one or both Fallopian tubes are distended to 10 millimeters or wider, **the Fallopian tube(s) should be removed or "clipped"** next to the uterus before the IVF treatment (see above).

- Continue on the next page -

From page

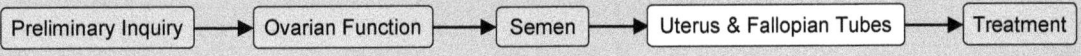

If any polyps, fibroids, or scarring are found on the sonohysterogram, HSG, or hysteroscopy, you will typically be scheduled for an operative hysteroscopy to have them removed.

Operative hysteroscopy is a simple outpatient procedure requiring only light anesthesia. A thin optical scope is passed through the cervical canal inside the uterus and the abnormal structure(s) are identified and removed. The recovery time after this procedure is typically very short.

If your uterine cavity was normal or once any abnormalities have been corrected, your fertility assessment is complete and you can proceed to *in vitro* fertilization (IVF) as your most meaningful path to a successful (pregnancy not ending in a miscarriage) conception.

To learn more about IVF treatment: **Go to page 89.**

---

[1] A protruding growth from the lining of the uterus.
[2] Benign growths of muscle cells that develop on the uterine wall.
[3] Outpatient procedure during which a thin optical scope is passed through the cervical canal inside the uterus and the endometrial cavity is visualized.
[4] Outpatient procedure during which a thin optical scope is inserted inside the abdomen through a small incision just below the bellybutton. It can be used to correct abnormalities of the reproductive organs.

Go to page

FERTILITY ASSESSMENT ALGORITHM

Preliminary Inquiry → Ovarian Function → Semen → Uterus & Fallopian Tubes → Treatment

ENDOMETRIAL CAVITY

Normally, you will be scheduled for a hysteroscopy to remove the cause of the filling defect. Hysteroscopy is a quick outpatient procedure requiring only very light anesthesia. A thin optical scope is passed through the cervical canal inside the uterus and the endometrial cavity is visualized. The cause of the filling defect is identified and removed. The recovery time after this procedure is typically very short.

The hysteroscopy may also find that your uterine cavity is normal and that the HSG's "abnormal" findings were an artifact (error).

Once your uterine cavity is normal, your fertility assessment is complete and you can proceed to *in vitro* fertilization treatment: **Go to page 89.**

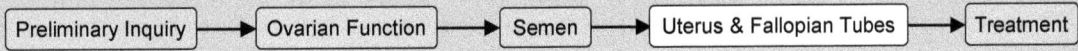

ENDOMETRIAL CAVITY

It is important to confirm the absence of polyps[1], scarring, and fibroids (myomas)[2] *inside* the uterus. The presence of any of these structures could reduce the probability of embryo implantation. On the other hand, they should not be considered as the sole cause of infertility. After their removal, the recommended treatment is the same as if the endometrial cavity had been normal.

Sonohysterogram, hysterosalpingogram (see below), or hysteroscopy[3] can be used to assess the endometrial cavity. Only *one* of these tests is necessary. If you have had *any* of them within the last six months, you do not need another one; otherwise you should have either a sonohysterogram or hysterosalpingogram.

- Sonohysterogram (a transvaginal ultrasound of the uterus) is the least invasive and least expensive of the above tests. It is typically done by your gynecologist. It can be done on any cycle day when you are not having your period. The ideal time for this test is just after the end of a menstrual period.

  A thin, soft catheter is passed through the cervix inside the uterus and a small amount of sterile saline solution is instilled. This allows the contours of the endometrial cavity to be visualized.

  If the ultrasound shows that one or both Fallopian tubes are distended to 10 millimeters or wider, **the Fallopian tube(s) should be removed or "clipped"** next to the uterus before the oocyte donation treatment. This severs the communication between the inside of the Fallopian tube(s) and the inside of the uterus. The secretion produced by the inner lining of the distended Fallopian tubes is embryo-toxic and could impede embryo implantation. You will need a laparoscopy[4] to have the Fallopian tube(s) removed or clipped.

- Hysterosalpingogram (HSG) is an x-ray of the uterus and Fallopian tubes. It is a quick procedure done in a radiology department.

  During the 10 to 20 minute test, a thin tube is passed through the cervical canal inside the uterine (endometrial) cavity and radio-opaque dye is slowly instilled. Two or more x-ray pictures are taken. You may experience menstrual-like cramping when the tube is being passed and when the dye is instilled. The result is available immediately and can be interpreted right away.

  If the HSG showed that one or both Fallopian tubes are distended to 10 millimeters or wider, **the Fallopian tube(s) should be removed or "clipped"** next to the uterus before the oocyte donation treatment (see above).

- Continue on the next page -

From page

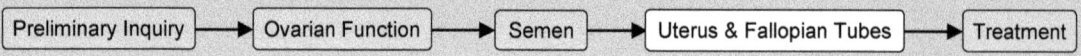

If any polyps, fibroids, or scarring are found on the sonohysterogram, HSG, or hysteroscopy, you will typically be scheduled for an operative hysteroscopy to have them removed.

Operative hysteroscopy is a simple outpatient procedure requiring only light anesthesia. A thin optical scope is passed through the cervical canal inside the uterus and the abnormal structure(s) are identified and removed. The recovery time after this procedure is typically very short.

If your uterine cavity was normal or once any abnormalities have been corrected, your fertility assessment is complete and you can proceed to oocyte donation treatment as your most meaningful path to a successful (pregnancy not ending in a miscarriage) conception.

To learn more about oocyte donation treatment: **Go to page 92.**

---

[1] A protruding growth from the lining of the uterus.
[2] Benign growths of muscle cells that develop on the uterine wall.
[3] Outpatient procedure during which a thin optical scope is passed through the cervical canal inside the uterus and the endometrial cavity is visualized.
[4] Outpatient procedure during which a thin optical scope is inserted inside the abdomen through a small incision just below the bellybutton. It can be used to correct abnormalities of the reproductive organs.

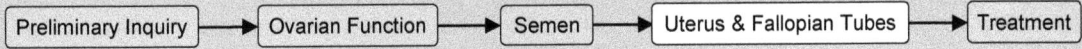

ENDOMETRIAL CAVITY

It is important to confirm the absence of polyps[1], scarring, and fibroids (myomas)[2] *inside* the uterus. The presence of any of these structures could reduce the probability of embryo implantation. On the other hand, they should not be considered as the sole cause of repeated miscarriages. After their removal, the recommended treatment is the same as if the endometrial cavity had been normal.

Sonohysterogram, hysterosalpingogram (see below), or hysteroscopy[3] can be used to assess the endometrial cavity. Only *one* of these tests is necessary. If you have had *any* of them within the last six months, you do not need another one; otherwise you should have either a sonohysterogram or hysterosalpingogram.

- Sonohysterogram (a transvaginal ultrasound of the uterus) is the least invasive and least expensive of the above tests. It is typically done by your gynecologist. It can be done on any cycle day when you are not having your period. The ideal time for this test is just after the end of a menstrual period.

  A thin, soft catheter is passed through the cervix inside the uterus and a small amount of sterile saline solution is instilled. This allows the contours of the endometrial cavity to be visualized.

- Hysterosalpingogram (HSG) is an x-ray of the uterus and Fallopian tubes. It is a quick procedure done in a radiology department.

  During the 10 to 20 minute test, a thin tube is passed through the cervical canal inside the uterine (endometrial) cavity and radio-opaque dye is slowly instilled. Two or more x-ray pictures are taken. You may experience menstrual-like cramping when the tube is being passed and when the dye is instilled. The result is available immediately and can be interpreted right away.

If any polyps, fibroids, or scarring are found on the sonohysterogram, HSG, or hysteroscopy, you will typically be scheduled for an operative hysteroscopy to have them removed.

Operative hysteroscopy is a simple outpatient procedure requiring only light anesthesia. A thin optical scope is passed through the cervical canal inside the uterus and the abnormal structure(s) are identified and removed. The recovery time after this procedure is typically very short.

- Continue on the next page -

From page

FERTILITY ASSESSMENT ALGORITHM

Preliminary Inquiry → Ovarian Function → Semen → Uterus & Fallopian Tubes → Treatment

If your uterine cavity was normal or once any abnormalities have been corrected, your fertility assessment is complete and you can proceed to a review of your recurrent miscarriage investigation: **Go to page 81.**

---

[1] A protruding growth from the lining of the uterus.
[2] Benign growths of muscle cells that develop on the uterine wall.
[3] Outpatient procedure during which a thin optical scope is passed through the cervical canal inside the uterus and the endometrial cavity is visualized.

Go to page

FERTILITY ASSESSMENT ALGORITHM

Preliminary Inquiry → Ovarian Function → Semen → Uterus & Fallopian Tubes → Treatment

RECURRENT MISCARRIAGES

The results of your fertility assessment have significantly reduced the possibility that your repeated (recurrent) miscarriages are caused by abnormal semen and egg quality or a uterine abnormality. You may decide to further investigate the possible causes of your repeated miscarriages with genetic, immunological, and additional hormonal tests. Such investigation is beyond the scope of this book. Your physician will recommend the necessary tests for you.

Approximately 50% of the time, even a very thorough investigation for recurrent miscarriages will not uncover their cause. The likely explanation is the fact that most miscarriages are due to suboptimal oocyte (egg) quality which may be very difficult to pinpoint. Many times this explanation is used after excluding all other possible contributing factors.

Most miscarriages are Mother Nature's way of preventing less-than-biologically-perfect embryos from proceeding to a live birth.

There is no direct treatment to improve the egg quality[1] but *in vitro* fertilization (IVF) has been successfully used for many couples with a history of recurrent miscarriages.

To learn more about *in vitro* fertilization: **Go to page 89.**

---

[1] Clinical evidence suggests that one's lifestyle may influence the quality of eggs being produced. See page 15 for lifestyle recommendations and try to adhere to them as closely as possible.

From page [ ]   Go to page [ ]

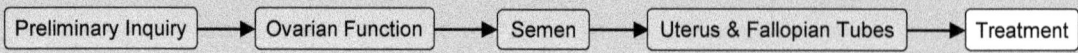

YOUR TREATMENT

Your fertility assessment did not show any *overt* cause of infertility. You may have had a "filling defect" in your endometrial cavity which has been corrected or, if you needed a laparoscopy, it found only endometriosis[1], or external myomas (fibroids)[2]. Approximately 25% of all fertility investigations do not uncover the reason for infertility.

Your test results provide you with the widest selection of treatment options:

1. Ovarian stimulation with oral medications and intercourse or intrauterine insemination (IUI)
2. Ovarian stimulation with injectable medications and intrauterine insemination
3. *In vitro* fertilization (IVF)

1. Ovarian stimulation with oral medications is the simplest, least expensive treatment that could increase your probability of conception. In this treatment, oral medications are taken for five days each cycle and your ovaries should respond by producing two or more eggs thereby increasing your pregnancy probability. To find out more about this treatment: **Go to page 85.**

2. Ovarian stimulation with injectable medications (FSH-follicle stimulating hormone, LH-luteinizing hormone) could substantially increase your probability of conception. This stimulation is commonly combined with intrauterine insemination (IUI). For some couples, this treatment could have an unacceptably high probability of a multiple pregnancy. To find out more about ovarian stimulation with injectable medications: **Go to page 87.**

3. If you decide to maximize your probability of having a live birth and, at the same time, minimize the risk of a multiple pregnancy, you should consider *in vitro* fertilization. This is the most involved and most expensive of these three treatment options but it gives you and your partner significantly more control over the outcome than the other two treatments. To find out more about *in vitro* fertilization: **Go to page 89.**

The sequence of the above treatments, from the simplest to the most complex (and most effective) does not imply that ovarian stimulation with oral medications is a prerequisite for ovarian stimulation with injectable medications or that treatment with injectable medications is a prerequisite for *in vitro* fertilization.

- Continue on the next page -

From page

FERTILITY ASSESSMENT ALGORITHM

Preliminary Inquiry → Ovarian Function → Semen → Uterus & Fallopian Tubes → Treatment

You could choose to proceed in this sequence hoping to minimize the cost of your infertility treatment, but you may also decide to go directly to ovarian stimulation with injectable medications or *in vitro* fertilization to increase your probability of pregnancy.

---

[1] The presence of tissue that normally grows inside the uterus in an abnormal anatomical location.
[2] Benign growths of muscle cells that develop on the outside uterine wall.

Go to page

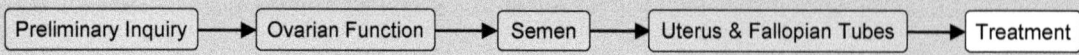

TREATMENT FOR HYPOTHALAMIC HYPOGONADISM

In hypothalamic hypogonadism, the pituitary gland does not produce adequate amounts of follicle stimulating hormone (FSH) and luteinizing hormone (LH), resulting in a lack of ovulation.

If increasing one's body weight and/or eliminating strenuous exercise and high levels of stress did not result in a resumption of ovulation and regular menstrual periods, ovarian stimulation with FSH and LH (injectable medications) can be a very effective and successful treatment.

FSH and LH ovarian stimulation is an involved and somewhat expensive treatment requiring close monitoring. Most gynecologists do not routinely perform FSH ovarian stimulations and you may need to be treated by an infertility specialist. The American Society for Reproductive Medicine (asrm.org) can be a valuable source for locating an infertility specialist in your area.

FSH and LH are administered daily by subcutaneous injections (under-the-skin) for approximately 7 to 12 days. The treatment is monitored with ultrasound examinations and blood measurements of estradiol (estrogen, E_2) levels. One or more eggs should develop within the ovaries.

Once the eggs are sufficiently mature, a subcutaneous injection of human chorionic gonadotropin (HCG) hormone triggers the final stages of maturation. Intercourse or intrauterine insemination is timed 20 to 40 hours after the HCG injection.

You should expect pregnancy probability per cycle of treatment up to 40% depending primarily on the female partner's age. Since FSH/LH ovarian stimulation typically results in the development of multiple eggs, it may not be possible to reliably limit the number of embryos (fertilized eggs) that implant.

The prospect of a high-order multiple pregnancy (triplets or more) can be the most significant "side-effect" of FSH/LH ovarian stimulation. For this reason, some couples will choose *in vitro* fertilization[1] (in which they can control the number of implanting embryos) over FSH/LH ovarian stimulation treatment.

A detailed description of FSH/LH ovarian stimulation can be found on page 103.

---

[1] See page 89 for a description of *in vitro* fertilization.

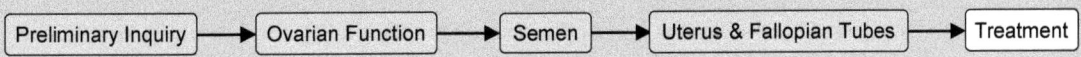

OVARIAN STIMULATION WITH ORAL MEDICATIONS

Ovarian stimulation with oral medications is a relatively inexpensive and simple infertility treatment. You should not need an infertility specialist for ovarian stimulation with oral medications, as most gynecologists are well versed in this type of ovarian stimulation.

The treatment starts on cycle day 3 or 5 after the onset of your menstrual period. This could be a spontaneous period or one induced with hormonal medications. Oral medications for ovarian stimulation are commonly given for 5 days (cycle day 3 through 7 or day 5 through 9). The ovaries should respond by producing one or more eggs and ovulation will typically take place around cycle day 15.

You may decide to have intercourse at the time of ovulation or your physician may recommend intrauterine insemination (IUI) with the male partner's semen (or donor semen if the male partner cannot produce sperm - page 59).

If you select to have IUI, the male partner collects a semen sample prior to the insemination. Sperm preparation for insemination is a multi-staged process of retrieving the highest quality sperm from the semen sample. It takes approximately one to two hours. The sperm are then loaded into an IUI catheter.

The thin, soft catheter is passed through the cervical canal to the top of the endometrial cavity and the sperm are gently released. The sperm quickly move into the Fallopian tubes and, hopefully, fertilize the egg(s).

Most couples should expect 5% to 15% probability of conception per cycle of ovarian stimulation with oral medications.

If you do not conceive, your period should start around cycle day 29. If your cycle is shorter than 26 or longer than 34 days, you likely did not ovulate. You could try another cycle of ovarian stimulation with oral medications but if your cycle is again less than 26 or more than 34 days, you should proceed to ovarian stimulation with injectable medications (page 87) or to *in vitro* fertilization (page 89).

If you did not conceive and your cycle was between 26 and 34 days, you could decide to repeat the ovarian stimulation up to 6 to 9 times. Each time you do not conceive the probability of pregnancy decreases with the subsequent treatment cycle. After 6 to 9 treatment cycles, the probability of pregnancy becomes too low to continue the treatment.

- Continue on the next page -

FERTILITY ASSESSMENT ALGORITHM

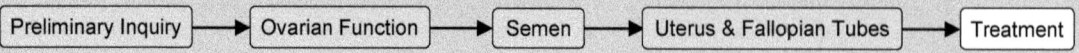

If you do not become pregnant using ovarian stimulation with oral medications, you should consider ovarian stimulation with injectable medications (page 87), or *in vitro* fertilization (page 89) as your next step.

It is important to note that intrauterine inseminations have not been shown to be useful in the treatment of male infertility. In fact, with the presence of male factor infertility there seems to be no higher probability of conception with artificial insemination than there is with intercourse. Couples with male factor infertility will need to consider *in vitro* fertilization treatment (page 89).

FERTILITY ASSESSMENT ALGORITHM

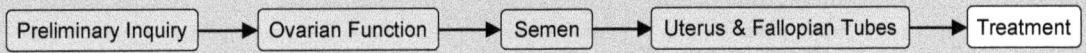

OVARIAN STIMULATION WITH INJECTABLE MEDICATIONS

Follicle stimulating hormone (FSH) secretion by the pituitary gland normally results in production of a single egg by the ovaries each cycle. It is possible to supplement the endogenous (by the pituitary gland) production of FSH with injections of FSH. This type of ovarian stimulation should result in the development of two or more eggs within the ovaries. The increased egg production, in turn, can improve the probability of pregnancy.

FSH ovarian stimulation is an involved and somewhat expensive treatment requiring close monitoring. Most gynecologists do not routinely perform FSH ovarian stimulations and you may need to be treated by an infertility specialist. The American Society for Reproductive Medicine (asrm.org) can be a valuable source for locating an infertility specialist in your area.

FSH is administered daily by subcutaneous (under-the-skin) injections for approximately 7 to 12 days. The treatment is monitored with ultrasound examinations and blood measurements of estradiol (estrogen, E_2) levels. Typically, two to four ultrasounds and blood draws are needed during a cycle of FSH ovarian stimulation.

Once the eggs are sufficiently mature, a subcutaneous injection of human chorionic gonadotropin (HCG) hormone triggers the final stages of maturation. Intercourse or intrauterine insemination is timed 20 to 40 hours after the HCG injection.

If you select to have IUI, the male partner collects a semen sample prior to the insemination. Sperm preparation for insemination is a multi-staged process of retrieving the highest quality sperm from the semen sample. It takes approximately one to two hours. The sperm are then loaded into an IUI catheter.

The thin, soft catheter is passed through the cervical canal to the top of the endometrial cavity and the sperm are gently released. The sperm quickly move into the Fallopian tubes and, hopefully, fertilize the egg(s).

You should expect the pregnancy probability per treatment cycle to be in the 10% to 20% range depending primarily on the female partner's age. Since FSH ovarian stimulation typically results in the development of multiple eggs, it may not be possible to reliably limit the number of embryos (fertilized eggs) that implant.

The prospect of a high-order multiple pregnancy (triplets or more) can be the most significant "side-effect" of FSH ovarian stimulation. For this reason, some couples will choose *in vitro* fertilization (in which they can control the number of implanting embryos) over FSH ovarian stimulation treatment.

- Continue on the next page -

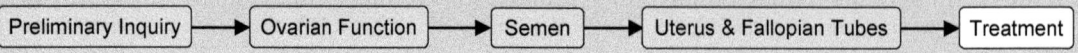

If you do not succeed, you could decide to repeat the ovarian stimulation up to 3 to 6 times. Each time you do not conceive the probability of pregnancy decreases with the subsequent treatment cycle. After 3 to 6 treatment cycles, the probability of pregnancy becomes too low to continue the treatment.

If you do not conceive with FSH ovarian stimulation, you should consider *in vitro* fertilization (page 89) as your next step.

It is important to note that intrauterine inseminations have not been shown to be useful in the treatment of male factor infertility. In fact, with the presence of male factor infertility there seems to be no higher probability of conception with artificial insemination than there is with intercourse. Couples with male factor infertility will need to consider *in vitro* fertilization treatment (page 89).

A detailed description of FSH ovarian stimulation can be found on page 103.

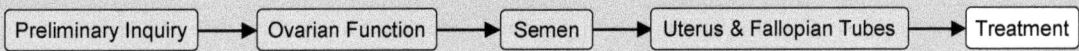

IN VITRO FERTILIZATION

In vitro fertilization (IVF) is one of the most effective treatments available to achieve a pregnancy. Most couples can expect a 20% to 45% probability of a live birth per IVF procedure. There are number of indications for choosing IVF:

- Traditionally IVF has been used to treat patients with a history of tubal blockage and pelvic adhesions (scarring).
- Male factor infertility is another common indication for IVF.
- Infertility associated with endometriosis and idiopathic (unexplained) infertility has been successfully overcome with IVF.
- Longstanding infertility (two years or longer) is another frequent reason for couples to choose IVF.
- Decreasing egg quality due to advanced female age may leave a couple with IVF as the only meaningful treatment option.
- IVF has been used to treat couples with recurrent miscarriages.
- Family gender balancing/pre-implantation genetic diagnosis (PGD)[1] can be added to IVF.
- If sperm must be aspirated from the testes or epididymis (convoluted tube, part of the spermatic duct system), IVF is required for fertilization of the eggs.

In vitro fertilization is performed in highly specialized IVF centers. The American Society for Reproductive Medicine (asrm.org) can be a valuable source for locating an *in vitro* fertilization clinic in your area.

IVF consists of the following steps:

1. Ovarian stimulation

   To maximize the probability of a live birth, follicle stimulating hormone (FSH), and possibly luteinizing hormone (LH), are used to stimulate production of as many high quality eggs as possible (usually 6 to 14 eggs). These hormones are administered as subcutaneous (under-the-skin) injections.

   During the 7 to 12 day ovarian stimulation, two or more ultrasound examinations and blood estradiol (estrogen, E_2) measurements are used to follow the development of the eggs. When the eggs are ready for retrieval, a subcutaneous injection of human chorionic gonadotropin (HCG) hormone is given. HCG completes the maturation process of the eggs.

   - Continue on the next page -

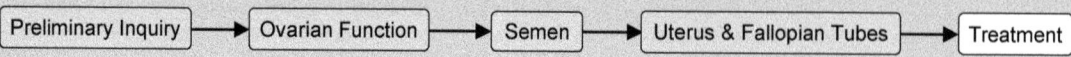

2. Egg retrieval

    Thirty-six hours after the HCG injection, a non-surgical oocyte retrieval is done. Using ultrasound guidance, a thin aspirating needle is passed through the top of the vagina into the ovarian follicles. Only the tip of the aspirating needle enters the pelvic area. Since the ovaries are located just above the top of the vagina, the tip of the needle is passed into the follicles without penetrating the uterus, cervix, or the Fallopian tubes.

3. Fertilization

    The male partner collects a semen sample by masturbation (unless sperm aspiration is needed[2]) and the highest quality sperm are added to the eggs six hours after the egg retrieval.

    If the fertility history suggests the possibility of male infertility significant enough to keep the eggs from being fertilized using regular laboratory methods, intracytoplasmic sperm injection (ICSI) is scheduled. ICSI is a micromanipulation technique in which a single sperm is inserted directly into an egg. ICSI is also done if sperm must be aspirated from the testes or epididymis (convoluted tube, part of the spermatic duct system).

    The next day, the eggs are examined for signs of fertilization. By the following day the fertilized eggs (embryos) reach 4 cells, 8 cells the day after, and by day five after egg retrieval they should reach the blastocyst (ready-to-hatch) stage.

    Embryos can be analyzed so that it is possible to conceive only from embryos of a desired gender (family gender balancing) or only from embryos that do not have a chromosomal abnormality for which they were analyzed (pre-implantation genetic diagnosis, PGD).

4. Embryo transfer

    One to five days after the egg retrieval, the resulting embryo(s) is/are transferred into the uterus by passing a thin embryo transfer catheter through the cervix to the top of the uterus.

    Extra embryos that are not transferred at this time can be cryopreserved (frozen) and stored in liquid nitrogen for potential future use.

5. Establishment of pregnancy

    A blood pregnancy test is scheduled approximately two weeks after the embryo transfer. A fetal heartbeat ultrasound is done two weeks after a positive pregnancy test.

    - Continue on the next page -

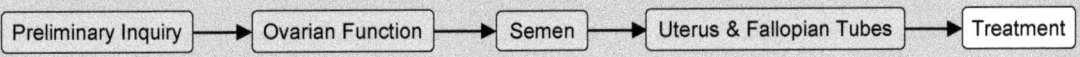

If you do not succeed in the first cycle of IVF, you could repeat the treatment. Or, if you have cryopreserved embryos, you may have another chance of a pregnancy from your embryos that have already been created.

In vitro fertilization can be a very accurate "test" for the egg and sperm quality. If you need more than one cycle of IVF, the subsequent cycles of treatment would be adjusted according to the knowledge gained, hopefully increasing your probability of success.

A more detailed description of *in vitro* fertilization can be found on page 105.

---

[1] See page 121 for a description of the PGD procedure.
[2] See page 120 for a description of the sperm aspiration procedure.

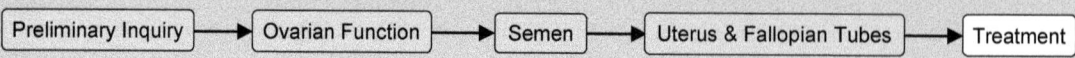

OOCYTE DONATION

The results of your infertility assessment indicate that you should/must consider oocyte (egg) donation as the most meaningful path to a successful conception. It is unlikely (but not totally impossible unless your ovaries have been removed) that you could successfully (pregnancy not ending in a miscarriage) conceive with your own eggs.

Oocyte donation is indicated for women who cannot or should not, for genetic reasons, become pregnant with their own eggs. It has been a part of infertility treatments since the 1980s and has become a "mainstream" infertility treatment. The most common reason for needing oocyte donation is age-related suboptimal oocyte quality.

One of the several advantages of oocyte donation treatment is its typically high probability of conception (50% to 75%). Another is the fact that it will be *you* getting pregnant, nourishing the baby throughout the pregnancy, and delivering the child. You may find that it matters little, if at all, from whose egg your pregnancy was conceived.

Since the uterus does not "grow old", you do not have to rush into oocyte donation treatment. It is important that you and your partner feel as comfortable as possible with the idea of conceiving with somebody else's eggs.

By the time the egg donation treatment is completed and you are ready for your prenatal care, the pregnancy has become indistinguishable from a spontaneous conception. It will be up to you and your partner to decide whether to share with your obstetrician that yours is an oocyte donation pregnancy.

Finding an oocyte donor is the first step in initiating the oocyte donation treatment. She can be a relative (blood-related to the embryo recipient), a friend, or you may use one of the many oocyte donor agencies available worldwide. Depending on the oocyte donor agency, you may be able to decide whether the oocyte donation treatment will be anonymous.

The oocyte donation procedure is similar to *in vitro* fertilization: after the egg donor's ovaries have been stimulated, the eggs are aspirated, inseminated with sperm from the embryo recipient's partner, incubated, and one or more of the resulting embryos are transferred into the recipient's uterus.

Oocyte donation is performed in highly specialized *in vitro* fertilization centers. The American Society for Reproductive Medicine (asrm.org) can be a valuable source for locating an IVF clinic offering oocyte donation treatment in your area.

- Continue on the next page -

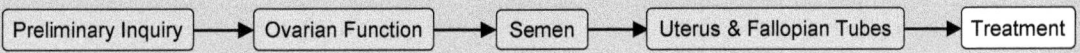

Oocyte donation consists of the following steps:

1. Ovarian stimulation

    To maximize the probability of a live birth, follicle stimulating hormone (FSH), and possibly luteinizing hormone (LH), are used to stimulate production of as many high quality eggs as possible (usually 6 to 14 eggs). These hormones are administered as subcutaneous (under-the-skin) injections.

    During the 7 to 12 day ovarian stimulation, two or more ultrasound examinations and blood estradiol (estrogen, E_2) measurements are used to follow the development of the eggs. When the eggs are ready for retrieval, your donor will take an injection of human chorionic gonadotropin (HCG) hormone. HCG completes the maturation process of the eggs.

2. Egg retrieval

    Thirty-six hours after the HCG injection, a non-surgical oocyte retrieval is done. Using ultrasound guidance, a thin aspirating needle is passed through the top of the vagina into the follicles. Only the tip of the aspirating needle enters the pelvic area. Since the ovaries are located just above the top of the vagina, the tip of the needle is passed into the follicles without penetrating the uterus, cervix, or the Fallopian tubes.

3. Fertilization

    The male partner collects a semen sample by masturbation (unless sperm aspiration is needed[1]) and the highest quality sperm are added to the eggs six hours after the egg retrieval.

    If the fertility history suggests the possibility of male infertility significant enough to keep the eggs from being fertilized using regular laboratory methods, intracytoplasmic sperm injection (ICSI) is scheduled. ICSI is a micromanipulation technique in which a single sperm is inserted directly into an egg. ICSI is also done if sperm must be aspirated from the testes or epididymis (convoluted tube, part of the spermatic duct system).

    The next day, the eggs are examined for signs of fertilization. By the following day the fertilized eggs (embryos) reach 4 cells, 8 cells the day after, and by day five after egg retrieval they should reach the blastocyst (ready-to-hatch) stage.

    - Continue on the next page -

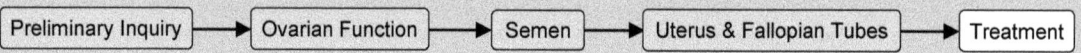

Embryos can be analyzed so that it is possible to conceive only from embryos of a desired gender (family gender balancing) or only from embryos that do not have a chromosomal abnormality for which they were analyzed (pre-implantation genetic diagnosis, PGD)[2].

4. Embryo transfer

One to five days after the egg retrieval, the resulting embryo(s) is/are transferred into the recipient's uterus by passing a thin embryo transfer catheter through the cervix to the top of the uterus.

The uterine lining has already been prepared to support the embryos by the use of estrogen and progesterone supplements. Extra embryos that are not transferred at this time can be cryopreserved (frozen) and stored in liquid nitrogen for potential future use.

5. Establishment of pregnancy

A blood pregnancy test is scheduled approximately two weeks after the embryo transfer. A fetal heartbeat ultrasound is done two weeks after a positive pregnancy test.

Estrogen and progesterone supplementation of the pregnancy continues for 6 to 8 weeks. By that time, the placenta produces enough of its own estrogen and progesterone so that the supplementation can stop. At that point, the pregnancy becomes indistinguishable from a conception through intercourse.

If you do not succeed in the first cycle of oocyte donation, you could repeat the treatment. Or, if you have cryopreserved embryos, you may have another chance of a pregnancy from your embryos that have already been created.

A more detailed description of oocyte donation can be found on page 108.

---

[1] See page 120 for a description of the sperm aspiration procedure.
[2] See page 121 for a description of the PGD procedure.

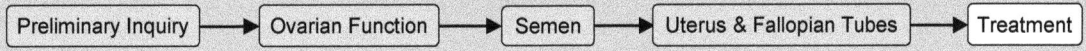

GESTATIONAL SURROGACY

The most common reasons for surrogacy are the intended mother's previous hysterectomy, congenital uterine malformations, or medical conditions that make pregnancy too risky for the intended mother.

Surrogacy is sometimes viewed as an alternative to adopting a child if the intended mother is unable to conceive or carry a pregnancy in her uterus. It allows a couple to have a genetically linked child.

There are two types of surrogate pregnancies:

- In traditional surrogacy, the surrogate is artificially inseminated with semen from the intended father. The surrogate provides the egg(s) and carries the pregnancy for the intended parents. The baby is genetically linked only to the intended father and not the intended mother. The insemination is typically a simple, inexpensive procedure not requiring an infertility specialist.

- In gestational surrogacy, the intended mother provides the egg(s), the intended father provides the sperm and *in vitro* fertilization is done to create embryos. The embryos are then transferred into the surrogate's uterus. The resulting baby is genetically unrelated to the surrogate. Most couples can expect a 20% to 45% probability of a live birth per gestational surrogacy procedure.

Finding a surrogate is the first step in initiating the gestational surrogacy treatment. Ideally, the surrogate should be a person emotionally invested in the pregnancy and the child's life (i.e., a close relative or a very close friend). A mature, emotionally stable surrogate carefully pre-screened by a surrogate agency can also be an excellent choice.

The gestational surrogacy procedure is similar to *in vitro* fertilization: after the intended mother's ovaries have been stimulated, the eggs are aspirated, inseminated with sperm from the intended father, incubated, and one or more of the resulting embryos are subsequently transferred into the surrogate's uterus.

Gestational surrogacy is performed in highly specialized *in vitro* fertilization centers. The American Society for Reproductive Medicine (asrm.org) can be a valuable source for locating an IVF clinic offering gestational surrogacy treatment in your area.

- Continue on the next page -

FERTILITY ASSESSMENT ALGORITHM

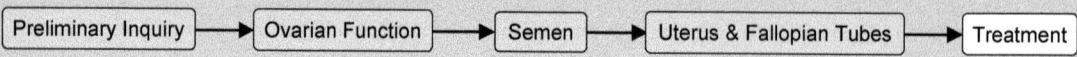

Gestational surrogacy consists of the following steps:

1. Ovarian stimulation of the intended mother

   To maximize the probability of a live birth, follicle stimulating hormone (FSH), and possibly luteinizing hormone (LH), are used to stimulate production of as many high quality eggs as possible (usually 6 to 14 eggs). These hormones are administered as subcutaneous (under-the-skin) injections.

   During the 7 to 12 day ovarian stimulation, two or more ultrasound examinations and blood estradiol (estrogen, E_2) measurements are used to follow the development of the eggs. When the eggs are ready for retrieval, the intended mother will take a subcutaneous injection of human chorionic gonadotropin (HCG) hormone. This hormone completes the maturation process of the eggs.

2. Egg retrieval

   Thirty-six hours after the HCG injection, a non-surgical oocyte retrieval is done. Using ultrasound guidance, a thin aspirating needle is passed through the top of the vagina into the ovarian follicles. Only the tip of the aspirating needle enters the pelvic area. Since the ovaries are located just above the top of the vagina, the tip of the needle is passed into the follicles without penetrating the uterus, cervix, or the Fallopian tubes.

3. Fertilization

   The intended father collects a semen sample by masturbation (unless sperm aspiration is needed[1]) and the highest quality sperm are added to the eggs six hours after the egg retrieval.

   If the fertility history suggests the possibility of male infertility significant enough to keep the eggs from being fertilized using regular laboratory methods, intracytoplasmic sperm injection (ICSI) is scheduled. ICSI is a micromanipulation technique in which a single sperm is inserted directly into an egg. ICSI is also done if sperm must be aspirated from the testes or epididymis (convoluted tube, part of the spermatic duct system).

   The next day, the eggs are examined for signs of fertilization. By the following day the fertilized eggs (embryos) reach 4 cells, 8 cells the day after, and by day five after egg retrieval they should reach the blastocyst (ready-to-hatch) stage.

   - Continue on the next page -

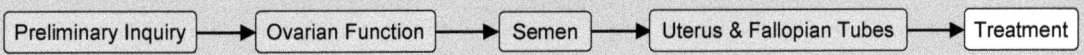

Embryos can be analyzed so that it is possible to conceive only from embryos of a desired gender (family gender balancing) or only from embryos that do not have a chromosomal abnormality for which they were analyzed (pre-implantation genetic diagnosis, PGD)[2].

4. Embryo transfer

   One to five days after the egg retrieval, the resulting embryo(s) is/are transferred into the surrogate's uterus by passing a thin embryo transfer catheter through the cervix to the top of the uterus.

   The uterine lining has already been prepared to support the embryos by the use of estrogen and progesterone supplements.

   Extra embryos that are not transferred at this time can be cryopreserved (frozen) and stored in liquid nitrogen for potential future use.

5. Establishment of pregnancy

   A blood pregnancy test is scheduled approximately two weeks after the embryo transfer. A fetal heartbeat ultrasound is done two weeks after a positive pregnancy test.

   Estrogen and progesterone supplementation of the pregnancy continues for 6 to 8 weeks. By that time, the placenta produces enough of its own estrogen and progesterone so that the supplementation can stop.

If you do not succeed in the first cycle of gestational surrogacy, you could repeat the treatment. Or, if you have cryopreserved embryos, you may have another chance of a pregnancy from your embryos that have already been created.

A more detailed description of gestational surrogacy can be found on page 112.

---

[1] See page 120 for a description of the sperm aspiration procedure.
[2] See page 121 for a description of the PGD procedure.

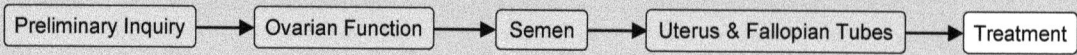

OOCYTE DONATION WITH GESTATIONAL SURROGACY

The results of your infertility assessment indicate that you should/must consider oocyte (egg) donation as the most meaningful path to a successful conception. It is unlikely (but not totally impossible unless your ovaries have been removed) that you could successfully conceive with your own eggs.

Oocyte donation is indicated for women who cannot or should not for genetic reasons become pregnant with their own eggs. The most common reason for needing oocyte donation is age-related suboptimal oocyte quality.

Gestational surrogacy is used if the intended mother is unable to conceive or carry a pregnancy in her uterus due to previous hysterectomy, congenital uterine malformations, or medical conditions that make pregnancy too risky for the intended mother.

Oocyte donation with gestational surrogacy allows a couple to have a child genetically linked to the intended father. It can be a very effective infertility treatment. The probability of a live birth depends greatly on the fertility potential of the oocyte donor and should be in the 50% to 75% range.

As an alternative to oocyte donation with gestational surrogacy treatment, you may want to consider traditional surrogacy. In traditional surrogacy, the surrogate is artificially inseminated with semen from the intended father. The surrogate provides both the egg(s) and carries the pregnancy for the intended parents. The insemination is typically a simple, inexpensive procedure not requiring an infertility specialist.

Finding an oocyte donor and a surrogate is the first step in initiating the oocyte donation with gestational surrogacy treatment. An egg donor can be a family member (blood-related to the intended mother) or you can use one of the many oocyte donor and surrogacy agencies. Ideally, the surrogate should be a person emotionally invested in the pregnancy and the child's life (i.e., a close relative or a very close friend). A mature, emotionally stable surrogate carefully pre-screened by a surrogate agency can also be an excellent choice.

The oocyte donation with gestational surrogacy procedure is similar to *in vitro* fertilization: after the egg donor's ovaries have been stimulated, the eggs are aspirated, inseminated with sperm from the intended father, incubated, and one or more of the resulting embryos are subsequently transferred into the surrogate's uterus.

- Continue on the next page -

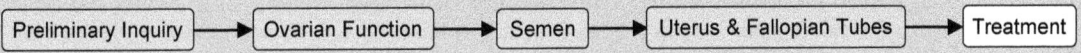

Oocyte donation with gestational surrogacy is performed in highly specialized *in vitro* fertilization centers. The American Society for Reproductive Medicine (asrm.org) can be a valuable source for locating an IVF clinic offering oocyte donation with gestational surrogacy treatment in your area.

Oocyte donation with gestational surrogacy consists of the following steps:

1. Ovarian Stimulation

    To maximize the probability of a live birth, follicle stimulating hormone (FSH), and possibly luteinizing hormone (LH), are used to stimulate production of as many high quality eggs as possible (usually 6 to 14 eggs). These hormones are administered as subcutaneous (under-the-skin) injections.

    During the 7 to 12 day ovarian stimulation, two or more ultrasound examinations and blood estradiol (estrogen, E_2) measurements are used to follow the development of the eggs. When the eggs are ready for retrieval, the donor will take an injection of human chorionic gonadotropin (HCG) hormone. HCG completes the maturation process of the eggs.

2. Egg Retrieval

    Thirty-six hours after the HCG injection, a non-surgical oocyte retrieval is done.

    Using ultrasound guidance, a thin aspirating needle is passed through the top of the vagina into the follicles. Only the tip of the aspirating needle enters the pelvic area. Since the ovaries are located just above the top of the vagina, the tip of the needle is passed into the follicles without penetrating the uterus, cervix, or the Fallopian tubes.

3. Fertilization

    The intended father collects a semen sample by masturbation (unless sperm aspiration is needed[1]) and the highest quality sperm are added to the eggs six hours after the egg retrieval.

    If the fertility history suggests the possibility of male infertility significant enough to keep the eggs from being fertilized using regular laboratory methods, intracytoplasmic sperm injection (ICSI) is scheduled. ICSI is a micromanipulation technique in which a single sperm is inserted directly into an egg. ICSI is also done if sperm must be aspirated from the testes or epididymis (convoluted tube, part of the spermatic duct system).

    - Continue on the next page -

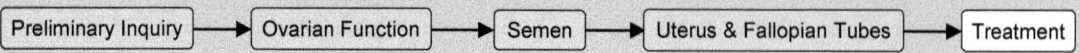

The next day, the eggs are examined for signs of fertilization. By the following day the fertilized eggs (embryos) reach 4 cells, 8 cells the day after, and by day five after egg retrieval they should reach the blastocyst (ready-to-hatch) stage.

Embryos can be analyzed so that it is possible to conceive only from embryos of a desired gender (family gender balancing) or only from embryos that do not have a chromosomal abnormality for which they were analyzed (pre-implantation genetic diagnosis, PGD)[2].

4. Embryo Transfer

One to five days after the egg retrieval, the resulting embryo(s) is/are transferred into the surrogate's uterus by passing a thin embryo transfer catheter through the cervix to the top of the uterus. The uterine lining has already been prepared to support the embryos by the use of estrogen and progesterone supplements.

Extra embryos that are not transferred at this time can be cryopreserved (frozen) and stored in liquid nitrogen for potential future use.

5. Establishment of Pregnancy

A blood pregnancy test is scheduled approximately two weeks after the embryo transfer. A fetal heartbeat ultrasound is done two weeks after a positive pregnancy test.

Estrogen and progesterone supplementation of the pregnancy continues for 6 to 8 weeks. By that time, the placenta produces enough of its own estrogen and progesterone so that the supplementation can stop.

If you do not succeed in the first cycle of oocyte donation with gestational surrogacy, you could repeat the treatment. Or, if you have cryopreserved embryos, you may have another chance of a pregnancy from your embryos that have already been created.

A more detailed description of oocyte donation with gestational surrogacy can be found on page 116.

---

[1] See page 120 for a description of the sperm aspiration procedure.
[2] See page 121 for a description of the PGD procedure.

SECTION IV
⌘
CONTEMPORARY REPRODUCTIVE TREATMENTS

Examples of Treatment Protocols

OVARIAN STIMULATION WITH INJECTABLE MEDICATIONS AND INTRAUTERINE INSEMINATION
(Treatment Protocol Example)

Ovarian stimulation with follicle stimulating hormone (FSH) is primarily indicated for women who do not have monthly menstrual periods and therefore do not ovulate. For these women, FSH ovarian stimulation has an exceptionally high success rate.

When FSH ovarian stimulation is used for women who are ovulatory and have monthly menstrual periods, it is hoped that by making their ovaries produce more than one egg, their probability of conception will be increased. This can hold true for some of the FSH ovarian stimulation candidates but overall the pregnancy rate is only in the 10% to 20% range per cycle of treatment.

To maximize the potential benefit of this treatment, FSH ovarian stimulation can be combined with one or two intrauterine inseminations (IUI).

FSH ovarian stimulation consists of:

1. Ovarian stimulation to induce growth of multiple eggs within the ovaries.
2. One or two intrauterine inseminations.
3. Establishment of pregnancy.

This is an example of a FSH ovarian stimulation treatment sequence. Actual treatment is individualized:

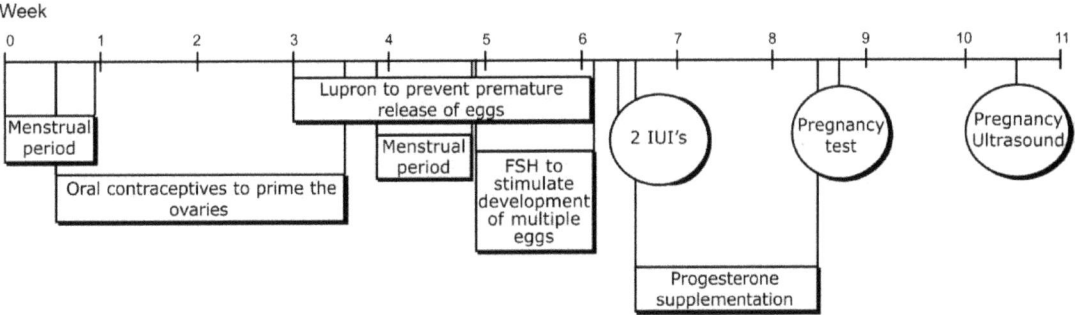

1. Ovarian stimulation

FSH ovarian stimulation treatment begins with the onset of a menstrual period. Oral contraceptives are started within the first seven days of the menstrual cycle. They prime the ovaries for an optimal response. Seven to ten days before the estimated onset of the next menstrual period, leuprolide (Lupron) injections begin. Leuprolide prevents premature release of the eggs from the ovaries prior to the intrauterine inseminations. The leuprolide injections are given subcutaneously (just under the skin). These injections are administered daily for approximately three to four weeks.

After seven to ten days of taking leuprolide, a menstrual period will start. Within two weeks of the onset of the period, FSH injections are added to the leuprolide. FSH stimulates maturation of multiple eggs in the ovaries. FSH injections, like leuprolide, are given subcutaneously with tiny needles. FSH injections are taken daily for approximately ten days.

During this time, the progress is monitored by ultrasound and estradiol (estrogen, E_2) blood levels. On the average, two to ten eggs will develop. Unlike *in vitro* fertilization, it is not possible to reliably control how many eggs develop, fertilize, and implant in the uterus. If a pregnancy occurs, there is a 25% to 30% probability of twins and an 8% to 25% risk that it will be a triplet or a higher order pregnancy.

Once the eggs are ready, the leuprolide and FSH are stopped and a single injection of human chorionic gonadotropin (HCG) hormone is taken. This is also a subcutaneous injection. This medication triggers the final stages of egg maturation. The next day, the first intrauterine insemination (IUI) is done. The second IUI is done the following day.

2. Intrauterine inseminations (IUI)

The male partner collects a semen specimen by masturbation the morning of the first intrauterine insemination. A second sample is collected the following morning. The highest quality sperm are extracted from the semen and are "loaded" into the tip of an IUI catheter. The catheter is then passed through the cervical canal to the top of the uterus and the sperm are gently released. The IUI usually takes a few seconds to complete. No resting is required afterwards.

3. Establishment of pregnancy

After the intrauterine insemination, the sperm rapidly move into the Fallopian tubes and await the arrival of the egg(s). The fertilization of the egg(s) takes place within the Fallopian tubes as it does with pregnancies conceived through intercourse. There is no need to restrict physical or sexual activity.

Within two to four days, the fertilized egg(s), now called embryo(s), will enter the uterus. The lining of the uterus has been made receptive for the embryo(s) through the action of ovarian hormones estrogen and progesterone. Ovarian progesterone production is supplemented with vaginal progesterone capsules or cream.

A blood pregnancy test can be done approximately two weeks after the IUI's. If the pregnancy test is positive, an ultrasound examination is scheduled two weeks later to visualize the implantation site and to look for a heartbeat within the embryo(s).

CONTEMPORARY REPRODUCTIVE TREATMENTS

IN VITRO FERTILIZATION
(Treatment Protocol Example)

In vitro fertilization (IVF) is one of the most effective treatments available to help infertile couples achieve pregnancy. Most couples will have a 25% to 45% probability of a live birth per IVF procedure. In addition to being a very powerful treatment for infertility, IVF is an excellent test of egg and sperm quality.

In vitro fertilization treatment consists of:

1. Ovarian stimulation to induce growth of multiple eggs within the ovaries.
2. Ultrasound guided egg retrieval.
3. Fertilization of the eggs.
4. Transfer of the resulting embryo(s) into the uterus.
5. Establishment of pregnancy.

This is an example of an IVF treatment sequence. Actual treatment is individualized:

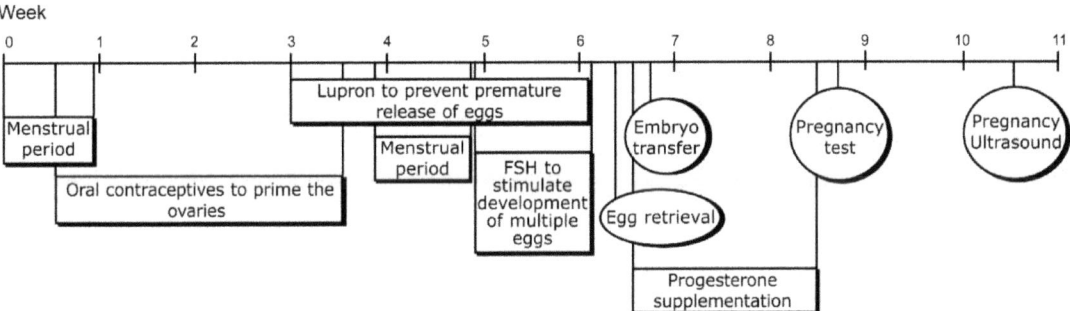

1. Ovarian stimulation

IVF treatment begins with the onset of a menstrual period. Oral contraceptives are started within the first seven days of the menstrual cycle. They prime the ovaries for an optimal response. Seven to ten days before the estimated onset of the next menstrual period, leuprolide (Lupron) injections begin. Leuprolide prevents premature release of the eggs from the ovaries prior to the egg retrieval procedure. The leuprolide injections are given subcutaneously (just under the skin). These injections are administered daily for approximately three to four weeks.

After seven to ten days of taking leuprolide, a menstrual period will start. Within two weeks of the onset of the period, follicle stimulating hormone (FSH) injections are added to the leuprolide. FSH stimulates maturation of multiple eggs in the ovaries. FSH injections, like leuprolide, are given subcutaneously with tiny needles. FSH injections are administered daily for approximately ten days.

During this time, the progress is monitored by ultrasound and estradiol (estrogen, E_2) blood levels. Once the eggs are ready, the leuprolide and FSH are stopped and a single injection of human chorionic gonadotropin (HCG) hormone is taken. This is also a subcutaneous injection. This medication triggers the final stages of egg maturation. Thirty-six hours after the HCG injection, the eggs are nonsurgically retrieved from the ovaries.

2. Ultrasound guided egg retrieval

Using ultrasound guidance, a tip of a thin needle is passed through the top of the vagina into the cul-de-sac (a space behind the uterus). The ovaries are located near the bottom of the cul-de-sac allowing the tip of the aspirating needle to enter the ovarian follicles and aspirate the follicular fluid from them. The fluid is examined under a microscope to identify the eggs. The egg retrieval takes approximately five to ten minutes. Medications are used for pain relief. It is possible to have a short lasting menstrual-like cramp sensation when the tip of the needle passes through the top of the vagina (once for each ovary). The actual follicle aspiration is typically not felt by the patient. The egg retrieval is a very safe procedure.

3. IVF laboratory

On average, eight to fourteen eggs are aspirated during the egg retrieval procedure. The eggs are identified under the microscope and are placed in culture medium filled petri dishes. The composition of the medium resembles the fluid secreted by the Fallopian tubes. This allows the eggs and embryos (fertilized eggs) to develop in the laboratory environment at the same rate as inside the Fallopian tubes.

The male partner collects a semen specimen by masturbation (unless sperm aspiration is needed) the day of the egg retrieval. The highest quality sperm are extracted from the semen and are combined with the eggs six hours after the egg retrieval. The process of fertilization takes place over a period of several hours during the night.

If the fertility history suggests a possibility of male infertility significant enough to keep the eggs from being fertilized this way, intracytoplasmic sperm injection (ICSI) is performed. In ICSI, a single sperm is inserted into an egg. This can significantly increase the fertilization rate for selected couples. ICSI is also done if sperm must be aspirated from the testes or epididymis (convoluted tube, part of the spermatic duct system).

Evidence of fertilization can be seen the next day, 14 to 16 hours after insemination. The fertilized eggs are transferred into growth medium and continue to be cultured in the IVF laboratory.

4. Embryo transfer

The embryo transfer is done one to five days after the egg retrieval. The embryo(s) is/are "loaded" into the tip of a very thin embryo transfer catheter in a very small volume of transfer medium. The catheter is then passed through the cervical canal to within 5 mm of the top of the uterine cavity and the embryo(s) are gently released. The transfer usually takes only a few seconds to complete. No resting is required afterwards.

The gamete embryologists assess the embryos prior to the embryo transfer to determine their likelihood of implantation. Most partners usually select two to four embryos for the transfer. Approximately one-third of IVF pregnancies are twins and there are very few triplet or higher order pregnancies.

There may be more embryos than the couple wish to have transferred. It is possible to cryopreserve (freeze) these embryos and store them in liquid nitrogen. The majority of the embryos should survive the cryopreservation and thawing process. The implantation rate of the surviving embryos is similar to the "fresh" embryos.

5. Establishment of pregnancy

After the embryo transfer, the front and back walls of the uterus gently hold the embryos, keeping them within the uterus. There is no restriction of physical or sexual activity.

The lining of the uterus is made receptive for the embryos through the action of the hormones estrogen and progesterone produced by the ovaries. Ovarian progesterone production is supplemented with vaginal progesterone capsules or cream.

A blood pregnancy test is done approximately two weeks after the embryo transfer. If the pregnancy test is positive, an ultrasound examination is scheduled two weeks later to visualize the implantation site and to look for a heartbeat. Once a heartbeat is seen, there is a 90% to 95% probability that the pregnancy will continue to a live birth. From that point on, the pregnancy becomes indistinguishable from a pregnancy conceived through intercourse.

OOCYTE DONATION
(Treatment Protocol Example)

Oocyte donation is the most effective treatment available to help infertile couples achieve pregnancy. It should provide 50% to 75% probability of a live birth per cycle of treatment. The procedure is similar to *in vitro* fertilization.

Oocyte donation consists of:

1. Ovarian stimulation to induce growth of multiple eggs within the egg donor's ovaries.
2. Ultrasound guided retrieval of the eggs.
3. Fertilization of the eggs with the male partner's semen.
4. Preparation of the embryo recipient's uterus for embryo transfer.
5. Embryo transfer.
6. Establishment of pregnancy.

This is an example of an oocyte donation treatment sequence. Actual treatment is individualized:

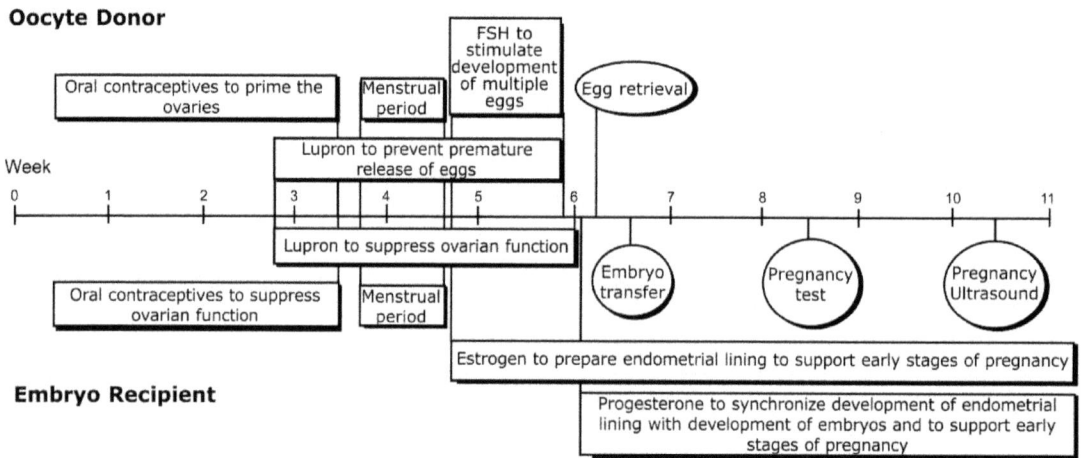

1. Ovarian stimulation

Oocyte donation treatment begins with the onset of the egg donor's menstrual period. Oral contraceptives are started within the first seven days of her menstrual cycle. They prime the ovaries for an optimal response. Seven to ten days before the estimated onset of her next menstrual period, leuprolide (Lupron) injections begin. Leuprolide prevents premature release of the eggs from the ovaries prior to the egg retrieval procedure. The leuprolide injections are given subcutaneously (just under the skin). They are administered daily for approximately three to four weeks.

After seven to ten days of taking leuprolide, the oocyte donor will start her menstrual period. Within two weeks of the onset of this period, she begins taking follicle stimulating hormone (FSH) injections in addition to the leuprolide.

FSH stimulates maturation of multiple eggs in the donor's ovaries. It is taken for approximately ten days. During this time her progress is monitored by ultrasound and estradiol (estrogen, E_2) blood levels.

Once the eggs are ready, she is instructed to stop taking the leuprolide and FSH and to take a single injection of human chorionic gonadotropin (HCG) hormone. This is also a subcutaneous injection. This medication triggers the final stages of egg maturation. Thirty-six hours after the HCG injection, the eggs are nonsurgically retrieved from the ovaries.

2. Ultrasound guided transvaginal egg retrieval

Using ultrasound guidance, a tip of a thin needle is passed through the top of the vagina and into the cul-de-sac (a space behind the uterus). The ovaries are located near the bottom of the cul-de-sac allowing the aspirating needle to enter the ovarian follicles and aspirate the follicular fluid from them. The fluid is examined under a microscope to identify the eggs.

The egg retrieval takes approximately five to ten minutes. Medications are used for pain relief. It is possible to feel a short lasting menstrual-like cramping sensation when the tip of the needle passes through the top of the vagina (once for each ovary). The actual follicle aspiration is typically not felt by the egg donor. The egg retrieval is a very safe procedure.

The egg retrieval procedure is the last step in the donor's participation. She will have her normal menstrual period within two weeks of the egg retrieval.

3. Laboratory

On average, eight to fourteen eggs are aspirated. The eggs are identified under the microscope and are placed into culture medium filled petri dishes. The composition of the medium resembles the fluid secreted by the Fallopian tubes. This allows the eggs and embryos (fertilized eggs) to develop in the laboratory environment at the same rate as inside the Fallopian tubes.

The male partner collects a semen specimen by masturbation (unless sperm aspiration is needed) the day of the egg retrieval. The highest quality sperm are extracted from the semen and are combined with the eggs six hours after the egg retrieval. The process of fertilization takes place over a period of several hours during the night.

If the fertility history suggests a possibility of male infertility significant enough to keep the eggs from being fertilized this way, intracytoplasmic sperm injection (ICSI) is performed. In ICSI, a single sperm is inserted into an egg. This can significantly increase the fertilization rate for selected couples. ICSI is also done if sperm must be aspirated from the testes or epididymis (convoluted tube, part of the spermatic duct system).

Evidence of fertilization can be seen the next day, 14 to 16 hours after insemination. The fertilized eggs are transferred into growth medium and continue to be cultured in the IVF laboratory.

4. Preparation of the female partner's uterus

The lining of the embryo recipient's uterus must be prepared to receive the embryos. The development of the uterine lining must be accurately synchronized with the development of the embryos. This is achieved by taking estrogen and progesterone.

If the embryo recipient has monthly menstrual periods, the treatment typically starts with taking oral contraceptives. They are used to suppress her ovarian function and to begin the process of synchronization. Oral contraceptives are started within the first seven days of the beginning of the menstrual cycle. Some embryo recipients do not need to take the oral contraceptives.

Seven to ten days before the estimated onset of the next menstrual period, leuprolide injections begin. Leuprolide "puts the ovaries to sleep" and temporarily stops their production of estrogen and progesterone. This estrogen and progesterone secretion by the ovaries would interfere with the development of the endometrial lining. The volume of the leuprolide injections is very small and they are given just under the skin. Women who do not have spontaneous menstrual periods typically do not have to take the leuprolide injections.

After seven to ten days of taking leuprolide the embryo recipient will start her menstrual period. Within one to three weeks of the onset of the period, she begins taking estrogen in the form of skin patches. The progress of the development of her uterine lining is monitored with ultrasound examinations and by estrogen blood levels. Once the donor is ready for the oocyte retrieval, the embryo recipient begins adding progesterone to the estrogen. The addition of progesterone opens the "window of receptivity" of her uterus and synchronizes development of its lining with the development of the embryos.

5. Embryo transfer

The embryo transfer is done one to five days after the egg retrieval. The embryo(s) is/are "loaded" into the tip of a very thin transfer catheter in a very small volume of transfer medium. The catheter is then passed through the cervical canal to within 5 mm of the top of the uterine cavity and the embryo(s) are gently released. The transfer usually takes a few seconds to complete. No resting is required afterwards and the embryo recipient can immediately resume her normal daily activities. She does not have to change her lifestyle as she goes through the oocyte donation treatment.

The gamete embryologists assess the embryos prior to the embryo transfer to determine their likelihood of implantation. Most partners usually select one to two embryos for the transfer.

Approximately one-third to one-half of oocyte donation pregnancies are twins and there are very few triplet or higher order pregnancies.

There may be more embryos than the couple wish to have transferred. It is possible to cryopreserve (freeze) these embryos and store them in liquid nitrogen. The majority of the embryos should survive the cryopreservation and thawing process. The implantation rate of the surviving embryos is similar to the "fresh" embryos.

6. Establishment of pregnancy

After the embryo transfer, the front and back walls of the uterus gently hold the embryos keeping them within the uterus. There is no need to restrict the embryo recipient's physical or sexual activity. She continues taking the estrogen patches and vaginal progesterone capsules or cream.

A blood pregnancy test is done approximately two weeks after the embryo transfer. If the pregnancy test is positive, an ultrasound examination is scheduled two weeks later to visualize the implantation site and to look for a heartbeat. Once a heartbeat is seen, there is a 90% to 95% probability that the pregnancy will continue to a live birth.

The embryo recipient continues to have her estrogen and progesterone blood levels monitored every one to two weeks. Six to eight weeks into the pregnancy the placenta produces enough of its own estrogen and progesterone that the supplementation can be discontinued. Once the female partner stops all her medications, the pregnancy becomes indistinguishable from a pregnancy conceived through intercourse.

GESTATIONAL SURROGACY
(Treatment Protocol Example)

In gestational surrogacy, the intended mother provides the eggs, the intended father provides the sperm and *in vitro* fertilization is done to create embryos. The embryos are then transferred into the surrogate's uterus.

Gestational surrogacy consists of:

1. Ovarian stimulation to induce growth of multiple eggs within the intended mother's ovaries.
2. Ultrasound guided retrieval of the eggs.
3. Fertilization of the eggs with the intended father's semen.
4. Preparation of the surrogate's uterus for embryo transfer.
5. Embryo transfer.
6. Establishment of pregnancy.

This is an example of a gestational surrogacy treatment sequence. Actual treatment is individualized:

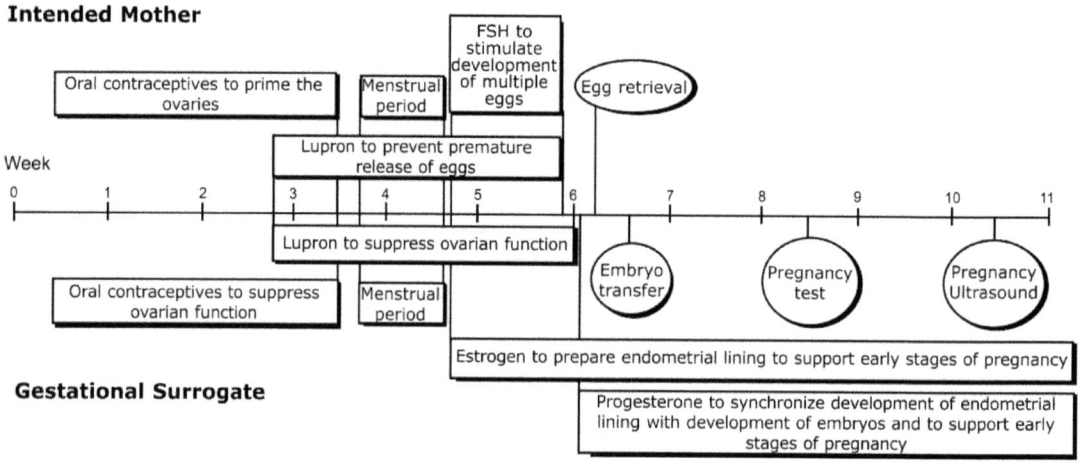

1. Ovarian stimulation

Gestational surrogacy treatment begins with the onset of the intended mother's menstrual period. Oral contraceptives are started within the first seven days of her menstrual cycle. They prime the ovaries for an optimal response. Seven to ten days before the estimated onset of the next menstrual period, leuprolide (Lupron) injections begin. Leuprolide prevents premature release of the eggs from the ovaries prior to the egg retrieval procedure. The leuprolide injections are given subcutaneously (just under the skin). They are administered daily for approximately three to four weeks.

After seven to ten days of taking leuprolide, the intended mother will start her menstrual period. Within one to two weeks of the onset of this period, she begins taking follicle stimulating hormone (FSH) injections in addition to the leuprolide.

FSH stimulates maturation of multiple eggs in the ovaries. The FSH medication is taken for approximately ten days. During this time her progress is monitored by ultrasound and estradiol (estrogen, E_2) blood levels.

Once the eggs are ready, she is instructed to stop taking the leuprolide and FSH and to take a single injection of human chorionic gonadotropin (HCG) hormone. This is also a subcutaneous injection. This medication triggers the final stages of egg maturation. Thirty-six hours after the HCG injection, the eggs are nonsurgically retrieved from the ovaries.

2. Ultrasound guided transvaginal egg retrieval

Using ultrasound guidance, a tip of a thin needle is passed through the top of the vagina and into the cul-de-sac (a space behind the uterus). The ovaries are located near the bottom of the cul-de-sac allowing the tip of the aspirating needle to enter the ovarian follicles and aspirate the follicular fluid from them. The fluid is examined under a microscope to identify the eggs.

The egg retrieval takes approximately five to ten minutes. Medications are used for pain relief. It is possible to feel a short lasting menstrual-like cramping sensation when the needle passes through the top of the vagina (once for each ovary). The actual follicle aspiration is typically not felt by the patient.

The egg retrieval is a very safe procedure. The egg retrieval procedure is the last step in the intended mother's participation in the treatment. She will have her normal menstrual period within two weeks of the egg retrieval.

3. Laboratory

On average, eight to fourteen eggs are aspirated. The eggs are identified under the microscope and are placed into culture medium filled petri dishes. The composition of the medium resembles the fluid secreted by the Fallopian tubes. This allows the eggs and embryos (fertilized eggs) to develop in the laboratory environment at the same rate as inside the Fallopian tubes.

The intended father collects a semen specimen by masturbation (unless sperm aspiration is needed) on the day of the egg retrieval. The highest quality sperm are extracted from the semen and are combined with the eggs six hours after the egg retrieval. The process of fertilization takes place over a period of several hours during the night.

If the fertility history suggests a possibility of male infertility significant enough to keep the eggs from being fertilized this way, intracytoplasmic sperm injection (ICSI) is performed. In ICSI, a single sperm is inserted into an egg. This can significantly increase the fertilization rate for selected couples. ICSI is also done if sperm must be aspirated from the testes or epididymis (convoluted tube, part of the spermatic duct system).

Evidence of fertilization can be seen the next day, 14 to 16 hours after insemination. The fertilized eggs are transferred into growth medium and continue to be cultured in the IVF laboratory.

4. Preparation of the surrogate's uterus

The lining of the surrogate's uterus must be prepared to receive the embryos. The development of the uterine lining must be accurately synchronized with the development of the embryos. This is achieved by taking estrogen and progesterone.

The surrogate's treatment typically starts with taking oral contraceptives. They are used to suppress her ovarian function and to begin the process of synchronization. Oral contraceptives are started within the first seven days of the menstrual cycle.

Seven to ten days before the estimated onset of the next menstrual period, leuprolide injections begin. Leuprolide "puts the surrogate's ovaries to sleep" and temporarily stops their production of estrogen and progesterone. This estrogen and progesterone secretion by the ovaries would interfere with the development of the endometrial lining.

After seven to ten days of taking leuprolide, the surrogate will have a menstrual period. Within one to three weeks of the onset of her period, she begins taking estrogen in the form of medicated skin patches. The progress of the development of her uterine lining is monitored with ultrasound examinations and by estrogen blood levels. Once the intended mother is ready for the oocyte retrieval, the surrogate begins adding progesterone to the estrogen. The addition of progesterone opens the "window of receptivity" for her uterus and synchronizes development of its lining with the development of the embryos.

5. Embryo transfer

The embryo transfer is done one to five days after the egg retrieval. The embryo(s) is/are "loaded" into the tip of a very thin transfer catheter in a very small volume of transfer medium. The catheter is then passed through the cervical canal to within 5 mm of the top of the uterine cavity and the embryos are gently released. The transfer usually takes a few seconds to complete. No resting is required afterwards and the surrogate can immediately resume her normal daily activities. She does not have to change her lifestyle as she goes through the surrogacy treatment.

The gamete embryologists assess the embryos prior to the embryo transfer to determine their likelihood of implantation. Most intended parents and surrogates usually select one to three embryos for the transfer. Approximately one-third of gestational surrogacy pregnancies are twins and there are very few triplet or higher order pregnancies.

There may be more embryos than the couple and the surrogate wish to have transferred. It is possible to cryopreserve (freeze) these embryos and store them in liquid nitrogen. The majority of the embryos should survive the cryopreservation and thawing process. The implantation rate of the surviving embryos is similar to the "fresh" embryos.

6. Establishment of pregnancy

After the embryo transfer, the front and back walls of the uterus gently hold the embryos keeping them within the uterus. There is no need to restrict surrogate's physical activity. She continues taking the estrogen patches and vaginal progesterone capsules or cream.

A blood pregnancy test is done approximately two weeks after the embryo transfer. If the pregnancy test is positive, an ultrasound examination is scheduled two weeks later to visualize the implantation site and to look for a heartbeat. Once a heartbeat is seen, there is a 90% to 95% probability that the pregnancy will continue to a live birth.

The surrogate continues to have her estrogen and progesterone blood levels monitored every one to two weeks. Six to eight weeks into the pregnancy the placenta produces enough of its own estrogen and progesterone that the supplementation can be discontinued.

OOCYTE DONATION WITH GESTATIONAL SURROGACY
(Treatment Protocol Example)

Oocyte donation with gestational surrogacy is one of the most effective treatments available to help infertile couples achieve pregnancy. Typically, there is a 50% to 75% probability of live birth per cycle of treatment. The procedure is similar to *in vitro* fertilization.

In this treatment, an egg donor provides the eggs, the intended father provides the sperm and *in vitro* fertilization is done to create embryos. The embryos are then transferred into the surrogate's uterus.

Oocyte donation with gestational surrogacy consists of:

1. Ovarian stimulation to induce growth of multiple eggs within the donor's ovaries.
2. Ultrasound guided retrieval of the eggs.
3. Fertilization of the eggs with the intended father's semen.
4. Preparation of the surrogate's uterus for embryo transfer.
5. Embryo transfer.
6. Establishment of pregnancy.

This is an example of an oocyte donation with gestational surrogacy treatment sequence. Actual treatment is individualized:

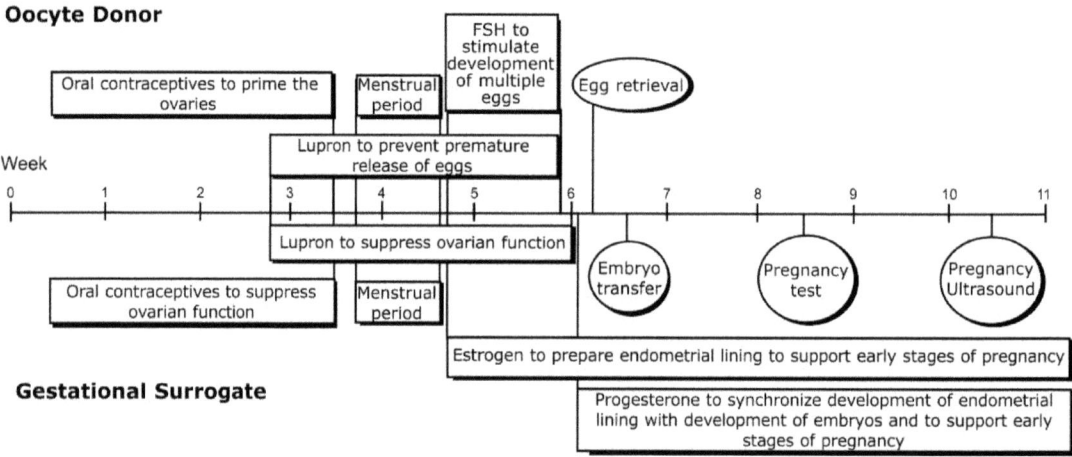

1. Ovarian stimulation

Oocyte donation with gestational surrogacy treatment begins with the onset of the egg donor's menstrual period. Oral contraceptives are started within the first seven days of her menstrual cycle. They prime the ovaries for an optimal response.

Seven to ten days before the estimated onset of the next menstrual period, leuprolide (Lupron) injections begin. Leuprolide prevents premature release of the eggs from the ovaries prior to the egg retrieval procedure. The leuprolide injections are given subcutaneously (just under the skin). These injections are administered daily for approximately three to four weeks.

After seven to ten days of taking leuprolide, the egg donor will start her menstrual period. Within two weeks of the onset of this period, she begins taking follicle stimulating hormone (FSH) injections in addition to the leuprolide.

FSH stimulates maturation of multiple eggs in the donor's ovaries. It is taken for approximately ten days. During this time her progress is monitored by ultrasound and estradiol (estrogen, E_2) blood levels.

Once the eggs are ready, she is instructed to stop taking the leuprolide and FSH and to take a single injection of human chorionic gonadotropin (HCG) hormone. This is also a subcutaneous injection. This medication triggers the final stages of egg maturation. Thirty-six hours after the HCG injection, the eggs are nonsurgically retrieved from the ovaries.

2. Ultrasound guided transvaginal egg retrieval

Using ultrasound guidance, a tip of a thin needle is passed through the top of the vagina and into the cul-de-sac (a space behind the uterus). The ovaries are located near the bottom of the cul-de-sac allowing the tip of the aspirating needle to enter the ovarian follicles and aspirate the follicular fluid from them. The fluid is examined under a microscope to identify the eggs.

The egg retrieval takes approximately five to ten minutes. Medications are used for pain relief. There may be a mild cramping sensation when the needle passes through the vaginal wall. The actual follicle aspiration is typically not felt by the egg donor. The egg retrieval is a very safe procedure. The egg retrieval procedure is the last step in the egg donor's participation in the treatment. She will have her normal menstrual period within two weeks of the egg retrieval.

3. Laboratory

On average, eight to fourteen eggs are aspirated. The eggs are identified under the microscope and are placed into culture medium filled petri dishes. The composition of the medium resembles the fluid secreted by the Fallopian tubes. This allows the eggs and embryos (fertilized eggs) to develop in the laboratory environment at the same rate as inside the Fallopian tubes.

The intended father collects a semen specimen by masturbation (unless sperm aspiration is needed) on the day of the egg retrieval. The highest quality sperm are extracted from the semen and are combined with the eggs six hours after the egg retrieval. The process of fertilization takes place over a period of several hours during the night.

If the fertility history suggests a possibility of male infertility significant enough to keep the eggs from being fertilized this way, intracytoplasmic sperm injection (ICSI) is performed. In ICSI, a single sperm is inserted into an egg. This can significantly increase the fertilization rate for selected couples. ICSI is also done if sperm must be aspirated from the testes or epididymis (convoluted tube, part of the spermatic duct system).

Evidence of fertilization can be seen the next day, 14 to 16 hours after insemination. The fertilized eggs are transferred into growth medium and continue to be cultured in the IVF laboratory.

4. Preparation of the surrogate's uterus

The lining of the surrogate's uterus must be prepared to receive the embryos. The development of the uterine lining must be accurately synchronized with the development of the embryos. This is achieved by taking estrogen and progesterone.

The surrogate's treatment typically starts with taking oral contraceptives. They are used to suppress her ovarian function and to begin the process of synchronization. Oral contraceptives are started within the first seven days of the menstrual cycle.

Seven to ten days before the estimated onset of the next menstrual period, leuprolide injections begin. Leuprolide "puts the surrogate's ovaries to sleep" and temporarily stops their production of estrogen and progesterone. This estrogen and progesterone secretion by the ovaries would interfere with the development of the endometrial lining.

After seven to ten days of taking leuprolide, the surrogate will have a menstrual period. Within one to three weeks of the onset of her period, she begins taking estrogen in the form of medicated skin patches. The progress of the development of her uterine lining is monitored with ultrasound examinations and by estrogen blood levels. Once the egg donor is ready for the oocyte retrieval, the surrogate begins adding progesterone to the estrogen. The addition of progesterone opens the "window of receptivity" for her uterus and synchronizes development of its lining with the development of the embryos.

5. Embryo transfer

The embryo transfer is done one to five days after the egg retrieval. The embryo(s) is/are "loaded" into the tip of a very thin transfer catheter in a very small volume of transfer medium. The catheter is then passed through the cervical canal to within 5 mm of the top of the uterine cavity and the embryos are gently released. The transfer usually takes a few seconds to complete. No resting is required afterwards and the surrogate can immediately resume her normal daily activities. She does not have to change her lifestyle as she goes through the surrogacy treatment.

The gamete embryologists assess the embryos prior to the embryo transfer to determine their likelihood of implantation. Most intended parents and surrogates usually select one to two embryos for the transfer. Approximately one-third to one-half of oocyte donation with gestational surrogacy pregnancies are twins and there are very few triplet or higher order pregnancies.

There may be more embryos than the couple and the surrogate wish to have transferred. It is possible to cryopreserve (freeze) these embryos and store them in liquid nitrogen. The majority of the embryos should survive the cryopreservation and thawing process. The implantation rate of the surviving embryos is similar to the "fresh" embryos.

6. Establishment of pregnancy

After the embryo transfer, the front and back walls of the uterus gently hold the embryos keeping them within the uterus. There is no need to restrict surrogate's physical activity. She continues taking the estrogen patches and vaginal progesterone capsules or cream.

A blood pregnancy test is done approximately two weeks after the embryo transfer. If the pregnancy test is positive, an ultrasound examination is scheduled two weeks later to visualize the implantation site and to look for a heartbeat. Once a heartbeat is seen, there is a 90% to 95% probability that the pregnancy will continue to a live birth.

The surrogate continues to have her estrogen and progesterone blood levels monitored every one to two weeks. Six to eight weeks into the pregnancy the placenta produces enough of its own estrogen and progesterone that the supplementation can be discontinued.

TESTICULAR AND EPIDIDYMAL SPERM ASPIRATION
(Treatment Protocol Example)

If sperm cannot be obtained by masturbation, they can be aspirated from the testes or epididymis (convoluted tube, part of the spermatic duct system).

Testicular or epididymal sperm aspiration is a technique that can be added to *in vitro* fertilization, oocyte donation, and gestational surrogacy treatments.

The most common conditions requiring sperm aspiration are:

- Previous vasectomy (with or without an attempt at reversal).
- Congenital absence of vas deferens (the connecting tube between testes and penis is missing).
- Sperm concentration and/or sperm quality so low that no normal sperm can be found in the ejaculate.

Sperm aspiration is typically an uncomplicated, quick outpatient procedure requiring only a small amount of local anesthetic.

Once sperm are successfully aspirated, intracytoplasmic sperm injection (ICSI) procedure is used for fertilization of eggs obtained through *in vitro* fertilization, oocyte donation, or gestational surrogacy treatment. In ICSI, a single sperm is inserted into an egg under a microscope using micro-instruments.

The sperm aspiration can be scheduled for the morning of egg retrieval procedure and the sperm is used for the ICSI the same day. Any sperm left over can normally be cryopreserved (frozen) and stored for potential subsequent use.

Alternatively, the sperm aspiration can be done prior to the egg retrieval. The collected sperm are frozen and stored in liquid nitrogen. The sperm sample is then thawed the morning of the egg retrieval and used for ICSI fertilization.

FAMILY GENDER BALANCING
PRE-IMPLANTATION GENETIC DIAGNOSIS
(Treatment Protocol Example)

Pre-implantation genetic diagnosis (PGD) is a technique that can be added to *in vitro* fertilization, oocyte donation, and gestational surrogacy treatments. With this method, the prospective parents will know the gender of each embryo with close to 100% accuracy prior to their transfer into the uterus.

This procedure can also identify genetically abnormal embryos. There are many incurable diseases or disorders which are genetically based, such as chromosome translocations, deletions, and inversions including Cystic Fibrosis, Fragile X, Myotonic Dystrophy, Thalassaemia, Tay-Sachs disease, and others. Using PGD it is possible to select only embryos that do not have a chromosomal abnormality for which they were analyzed.

PGD requires the creation of embryos by *in vitro* fertilization, oocyte donation, or gestational surrogacy treatment. It consists of:

1. Removal of embryonic cells.
2. Embryonic chromosomal analysis.
3. Extended embryo culture.

1. Removal of embryonic cells

Once embryos reach the five to eight cell stage (three days after egg retrieval), an embryo biopsy is performed by creating an opening in the egg shell around the embryo.

Since at this stage any cell inside the embryo has full developmental potential, it is possible to safely remove a single cell through this opening using a micropipette. The procedure is performed under a special microscope with micromanipulators.

2. Embryonic chromosomal analysis

After the biopsy, the embryos are placed back in an incubator to await the results of the genetic analysis. The biopsied cell's chromosomes are then analyzed.

3. Extended embryo culture

The genetic analysis takes approximately one to two days. Once the result of the PGD analysis is obtained, embryo(s) of the desired gender and/or that did not show chromosomal abnormalities for which they were analyzed is/are transferred. The embryos are typically transferred at the morula (day 4) or blastocyst (day 5) stage.

The accuracy of PGD cell analysis approaches 100%, but it is not guaranteed. It is possible, even though highly unlikely, that an embryo that has tested as normal may not be genetically perfect.

So far, there is no evidence that PGD embryos result in an increased probability of abnormalities in the baby or that the risk of birth defects is higher than the usual risk of abnormalities (2% to 4%) when compared to conceptions that occur spontaneously.

APPENDIX

APPENDIX

CALCIUM CHANNEL BLOCKERS

The main clinical usage of calcium channel blockers is to lower blood pressure. Calcium ion channel blocker medications may adversely affect sperm fertilizing potential.

If the male partner has to stay on calcium channel blocker medication, intracytoplasmic sperm injection (ICSI) procedure may be needed for fertilization of eggs. ICSI is a micromanipulation technique in which a single sperm is inserted directly into an egg. ICSI requires obtaining eggs through *in vitro* fertilization, oocyte donation, or gestational surrogacy treatment.

The most commonly used calcium channel blocker medications are:

- Amlodipine (Norvasc)
- Aranidipine (Sapresta)
- Azelnidipine (Calblock)
- Barnidipine (HypoCa)
- Benidipine (Coniel)
- Bepridil (Vascor)
- Cilnidipine (Atelec, Cinalong, Siscard)
- Clevidipine (Cleviprex)
- Diltiazem (Cardizem, Dilacor, Tiazac)
- Efonidipine (Landel)
- Felodipine (Plendil)
- Gallopamil (Procorum, D600)
- Lacidipine (Motens, Lacipil)
- Lercanidipine (Zanidip)
- Manidipine (Calslot, Madipine)
- Nicardipine (Cardene, Carden SR)
- Nifedipine (Procardia, Adalat)
- Nilvadipine (Nivadil)
- Nimodipine (Nimotop)
- Nisoldipine (Baymycard, Sular, Syscor)
- Nitrendipine (Cardif, Nitrepin, Baylotensin)
- Pranidipine (Acalas)
- Verapamil (Calan, Isoptin, Covera, Verelan)

ABOUT THE AUTHOR

FRANCIS POLANSKY, M.D.

Dr. Polansky grew up in Prague, Czech Republic. Soon after graduating from Charles University Medical School in Prague in 1978, he escaped from the communist regime of the former Czechoslovakia and came to the United States to begin his postgraduate training.

He started his residency in Obstetrics and Gynecology in 1979 and within few months, he fell in love with the intricacies and challenges of infertility diagnosis and treatment. He decided at that time to sub-specialize in Reproductive Endocrinology and Infertility (REI).

It was at that point that he realized that there was a lack of a unified scientific approach to infertility investigation. There seemed to be almost as many different ideas about infertility as there were physicians treating infertile couples. Wondering whether there was a way to logically organize infertility investigation, Dr. Polansky started to develop a framework of a systematic approach to infertility.

After finishing his OB/GYN residency at Stanford University in 1983, Dr. Polansky spent an additional two years at Stanford sub-specializing in REI. He finished his Reproductive Endocrinology and Infertility fellowship in 1985 and remained at Stanford University as a teaching faculty for another two years. During this time, together with Dr. Emmet Lamb, he co-founded the Stanford In Vitro Fertilization Clinic.

In the fall of 1987, he ventured out to start Nova IVF clinic and, subsequently, Bay IVF Center in Palo Alto, CA where he has continued to provide advanced reproductive treatments for his patients and education for physicians in OB/GYN residency training.

For more than 30 years, he has continued to refine his understanding of infertility trying to develop the most logical way of addressing the needs of infertile couples. The Fertility Assessment Algorithm™ in this book is the product of his research, teaching, and "hands-on" patient care.

Dr. Polansky was board certified in Obstetrics and Gynecology in 1986 and in Reproductive Endocrinology and Infertility in 1988. In 1986 he received the "Outstanding Professor of the Teaching Faculty, Department of Gynecology and Obstetrics, Stanford University" award.

INDEX

A

acid-alkaline balance, 15
acid-forming foods, 15
acupuncture, 17
anovulation, 7

B

baby for you, 1, 11
biological child, ix, 11

C

calcium channel blockers, 125
Chinese medicine, 17
conception, 5, 10, 11, 103
cost, 1
cryopreservation of embryos, 107, 111, 114, 119

E

ectopic pregnancy, 10, 11
egg quality, 6, 8, 11
egg retrieval, 105, 106, 108, 109, 110, 112, 113, 114, 117, 118, 120, 121
embryo, 6, 11, 104, 106, 107, 108, 110, 111, 112, 114, 115, 116, 118, 119, 121, 122
embryo transfer, 106, 108, 110, 112, 114, 116, 118
embryonic chromosomal analysis, 121
embryos, 9, 11, 105, 106, 107, 109, 110, 111, 112, 113, 114, 115, 116, 117, 118, 119, 121, 122
essential fatty acids, 16
estradiol, 21
estrogen, 6, 21, 104, 106, 107, 109, 110, 111, 113, 114, 115, 117, 118, 119
exercise, 16

F

Fallopian tubes, 5, 6, 10, 21, 104, 106, 109, 113, 117
family gender balancing, 121
female age, 8, 9, 11
female fertility, 8
Fertility Assessment Algorithm, ix, 21, 126
fertility investigation, ix, 1, 21
fertility potential, 5, 7, 8, 9, 11, 15, 17
fertilization, 5, 6, 9, 11, 104, 105, 106, 108, 109, 110, 112, 113, 114, 116, 117, 118, 120, 125, 126
fire retardant chemicals, 17
follicle, 6, 8, 21, 103, 105, 106, 109, 113, 117
FSH, 21, 103, 104, 106, 109, 113, 117
FSH ovarian stimulation, 103

G

gamete embryologists, 107, 110, 114, 118
genetic disorders, 9
gestational surrogacy, 112, 116, 120, 121, 125

H

HCG hormone, 104, 106, 109, 113, 117
hormonal production, 6, 8
hysterosalpingogram (HSG), 21
hysteroscopy, 21

I

implantation, 6, 9, 104, 107, 110, 111, 114, 115, 118, 119, 121
in vitro fertilization (IVF), 104, 105, 106, 107, 108, 110, 112, 114, 116, 118, 120, 121, 125, 126
inadequate hormonal production, 8

infertility, ix, 1, 5, 6, 7, 8, 9, 11, 105, 106, 109, 113, 118, 126
infertility investigation, 11, 126
insemination, 7, 10, 106, 110, 114, 118
intercourse, 5, 7, 10, 104, 107, 111
intracytoplasmic sperm injection (ICSI), 7, 120
intrauterine insemination (IUI), 103, 104

L

laparoscopy, 21
lifestyle, 8, 110, 114, 118
live birth, 5, 8, 10, 11, 15, 107, 108, 111, 115, 116, 119
loss of fertility, 9
Lupron, 103, 105, 108, 112, 117

M

male infertility, 6, 7
menopause, 8, 17
menstrual cycle, 103, 105, 108, 110, 112, 114, 116, 118
miscarriage, 9, 10, 17

O

oocyte donation, 10, 108, 110, 111, 116, 119, 120, 121, 125
optimize your fertility potential, 5
ovarian dysfunction, 8
ovarian follicles, 8, 106, 109, 113, 117
ovarian stimulation, 103, 105, 108, 112, 116
ovaries, 5, 8, 9, 21, 103, 104, 105, 106, 107, 108, 109, 110, 112, 113, 114, 116, 117, 118
ovulation, 5, 6, 7, 10

P

pelvic adhesions, 10
PGD, 121, 122
physical condition, 5, 15
pregnancy test, 6, 104, 107, 111, 115, 119

probability of a live birth, 8, 105
progesterone, 6, 8, 104, 107, 110, 111, 114, 115, 118, 119

R

receptivity of the mother's body, 11
reproductive treatments, 7, 9, 10, 11, 126
retrieval of eggs, 108, 112, 116
risk of chromosomal abnormality, 9

S

semen analysis, 21
smoking, 17
speed of conception, 5
sperm, ix, 6, 7, 10, 21, 104, 105, 106, 109, 112, 113, 116, 117, 118, 120, 125
sperm aspiration, 120
stress, 17
successful pregnancy, 3, 1, 5
surrogate, 112, 114, 115, 116, 118, 119

T

trans fatty acids, 16
tubal blockages, 10

U

ultrasound, 21, 104, 106, 107, 109, 110, 111, 113, 114, 115, 117, 118, 119
unexplained infertility, 7
uterine lining, 6, 11, 110, 114, 118
uterine receptivity, 11
uterus, 5, 6, 10, 11, 21, 75, 78, 80, 104, 105, 106, 107, 108, 109, 110, 111, 112, 113, 114, 115, 116, 117, 118, 119, 121

V

vasectomy, 7, 120
vitamins, 16
volatile organic compounds (VOC), 16

www.ingramcontent.com/pod-product-compliance
Ingram Content Group UK Ltd.
Pitfield, Milton Keynes, MK11 3LW, UK
UKHW051254180426
11947UKWH00020B/1718